LABYRINTH FISH

Helmut Pinter

LABYRINTH FISH

62 Color Photographs
24 Drawings
14 Maps

New York • London • Toronto • Sydney

Technical Adviser: Dr. Jürgen Lange
American Consulting Editor: Marshall Ostrow
Translator: Dieter Schulz
Ichthyology Consultant:
  Warren C. Freihofer
  Associate Curator, Dept. of Ichthyology
  California Academy of Sciences
Cover design: A. Krugman with a photograph by Helmut Pinter

First English language edition published in 1986 by Barron's Educational Series, Inc.

© 1984 Eugen Ulmer GmbH & Co., Stuttgart, Germany.

The title of the German edition is *Labyrinthfische*

All inquiries should be addressed to:
Barron's Educational Series, Inc.
250 Wireless Boulevard
Hauppauge, New York 11788

Library of Congress Catalog Card No. 86-3516
International Standard Book No. 0-8120-5635-3

Library of Congress Cataloging-in-Publication Data
Pinter, Helmut.
  Labyrinth fish.

  Translation of: Labyrinthfische.
  Includes index.
  1. Labyrinth fishes. I. Title.
SF458.L26P5613   1986   639.3′44   86-3516
ISBN 0-8120-5635-3

PRINTED IN HONG KONG

789   490   98765432

CONTENTS

PREFACE

During the last few decades of the nineteenth century, the aquarium hobby, then only in its infancy in the Western world, was measurably influenced by the early importations of labyrinth fishes. Since then labyrinth fishes have played a major role in the popularization of the aquarium hobby. This occurred despite the relatively few labyrinthid species compared with the speciose cyprinids of Asia, killies and cichlids of Africa, and characoids, cichlids, and live-bearers of Central and South America.

The early popularity of the labyrinth fishes of Southeast Asia probably occurred for several reasons, one being their magnificent finnage and color patterns. All the gouramis, for example, have long-based dorsal and anal fins that extend from the front part of the back and belly almost to the tail and long feelerlike pelvic fins. As far as colors are concerned, few fishes, marine or freshwater, are more colorful than the male dwarf gourami *(Colisa lalia),* for instance, which has alternating red and iridescent blue vertical stripes, or the pearl gourami *(Trichogaster leeri),* which has a fiery orange abdomen with an iridescent pearly pattern over the rest of its body. These are but a few examples of the garish colors of some of the more popular labyrinth fishes. These fishes also became popular because the majority are relatively easy to breed and many of them engage in interesting and easily observable spawning rituals and brood care. Furthermore, they were easy to import during a time when rapid air transportation was not yet invented. Exotic fishes from tropical and subtropical regions of the world had to withstand the rigors of lengthy sea voyages, often under the supervision of persons not properly trained to care for them. The unique air-breathing organs of labyrinth fishes allowed them to survive these perilous journeys in crowded vessels containing polluted, oxygen-poor water. Other kinds of fishes not having this unique organ rarely survived transoceanic shipment.

Labyrinth fishes are described in this book from as many aspects as possible, and because of their air-breathing similarities, a brief overview of the snakeheads *(Channa* sp.) is also given. Some of the details of the taxonomy and geographic distribution of certain species are sketchy at best and of necessity must remain open questions. There are several interesting reasons for this. Because of the evolutionary position of some species, some of their characteristics are not as clearly defined as they are in other species; thus positive definition or identification is difficult to achieve. There are also problems in the description of some of the distributional areas in Southeast Asia, which is where most labyrinth fishes occur. For example, some of the densest human populations in the world occur there. This has caused the faunal distribution to be disrupted by water pollution, overfishing of some species, and inadvertent or intentional release of nonindigenous species. Another problem with some labyrinth species as well as other kinds of fishes lies in their location in remote areas that are difficult, if not impossible, to explore, let alone enumerate the fauna. A further problem lies in the local politics of such countries as China, Kampuchea, Vietnam, and others, where it is often not possible to verify older taxonomic and geographic data.

Despite these difficulties and thanks to the valuable help of a number of knowledgeable, well-known labyrinth fish experts, this illustrated comprehensive overview of labyrinth fishes has come to fruition. I am especially grateful to Dr. Walter Foersch, who provided me with much important advice and contributed considerably to the illustration of this book. I also thank Dr.

Preface

Jurgen Lange for his kindness in reading the manuscript and Dietrich Schaller, whose detailed knowledge of Southeast Asian fish fauna provided valuable information on the distribution of many of the labyrinth fishes discussed herein. My gratitude also goes to Jörg Hess, Lothar Seegers, and Sune Holm, who were kind enough to make many photographs available.

It is my hope that this book contributes to the knowledge of keeping the old familiar labyrinth fishes and stimulates interest in the newly discovered species.

Helmut Pinter

8

INTRODUCTION

FISHES WITH ACCESSORY BREATHING ORGANS

In some tropical zones, freshwater fishes inhabit waters in which the oxygen content is temporarily or permanently so low that the fishes' oxygen requirements cannot be supplied by means of gills alone. Aside from lungfishes and the lobe-finned polypterids, which take in oxygen through their lungs or a lunglike air bladder, a number of other kinds of fishes have various means of utilizing atmospheric oxygen and thus easily adapt to oxygen-poor waters. These additional means of oxygen intake supplement gill respiration and can even replace it almost completely for varying periods of time, depending on conditions and species. Fishes absorb oxygen dissolved in water as the water passes over the gills, but fishes with accessory breathing organs or oxygen-absorbing surfaces also take in air at the water surface. The oxygen contained therein is absorbed by the accessory breathing apparatus and is thereby made available to the fish.

The additional breathing apparatus in some fishes consists of mucous membranes in the buccal cavity (mouth) or in the intestines or air bladder. The supplementary breathing apparatus in catfish of the genus *Clarius,* for example, is particularly effective. It consists of saclike enlargements of the gill cavities, which contain large vascularized surface areas suitable for oxygen intake. This special breathing apparatus permits these fishes to migrate overland from one body of water to another when adverse conditions warrant such migration. Some fishes have a duct connecting the air bladder to the alimentary tract. In these fishes, the heavily vascularized inner surface of the air bladder is capable of absorbing large quantities of oxygen from air that is swallowed and passed into it via the duct. This form of supplementary breathing is found not only in exotic tropical fishes, such as *Erythrinus* species and *Gymnarchus* species, but also in temperate zone fishes, such as the dogfish (*Amia calva*) and mudminnows of the genus *Umbra.*

Supplementary breathing organs are also generally used when oxygen saturation in the water is normal. This suggests that such organs probably evolved simultaneously with a secondary degeneration of the gill-breathing apparatus, so that these fishes cannot supply their complete oxygen needs by way of gills alone. Armored catfish of the genera *Corydoras* and *Callichthys,* for example, regularly dash to the surface for a bubble of air, whether they are in oxygen-poor or oxygen-rich water. Some fishes with additional breathing organs will die if they are prevented from reaching the surface. This phenomenon has been examined in detail in only a few species and is not generally so in fishes that use accessory breathing organs only in emergencies. One group of fishes that uses both gill-breathing and air-breathing systems regularly and almost simultaneously is the labyrinth fishes (anabantoids), of which numerous species are widely dispersed throughout Southeast Asia and some in Africa.

WHAT IS A LABYRINTH FISH?

Before we deal with the characteristics of labyrinth fishes more closely, it is necessary for the sake of clarity to explain a linguistic ambiguity. Almost all vertebrates have a labyrinth as part of the inner ear. This labyrinth serves as the organ of equilibrium and is so named because of its complex structure. Not only do mammals, birds, reptiles, and amphibians have such organs, but even bony fishes are equipped with them. This equilibrium organ has nothing to do with the air-intake organ of labyrinth fishes, which because of its highly convoluted, multi-chambered inner structure is also called a labyrinth.

The additional breathing organs of labyrinth fishes are located in the gill cavities, which in these fishes are somewhat expanded anteriorly and posteriorly as well as upward into the skull. The gill cover is partially connected to the first gill arch (branchial bone), and the first and second gill arches are grown together in such a way that a passage is formed from the mouth through the gill cavity and into the labyrinth cavity, so that when air is gulped by these fishes it passes directly into the labyrinth cavity. In larval fish an upward bony projection is formed from the top of the first gill arch. This projection is called the epibranchial bone. In the course of development the epibranchial bone grows upward into the labyrinth cavity and terminates in the highly convoluted supra-branchial labyrinth organ,

which is comprised of numerous heavily vascularized lamellae. The development of the labyrinth organ is, however, gradual. Most larval labyrinth fishes at first breathe exclusively with their gills.

Recent investigations have shown that air intake from the surface does not occur in the same manner in all genera. In the *Ctenopoma* species, for example, the used air is pushed out through the mouth by water coming in through the opened gill covers. Thus the labyrinth cavity is filled with water before the fish takes in fresh air. In the closely related Indian climbing perch (*Anabas testudineus*), no water is taken into the labyrinth cavity. In this species the fresh air taken in through the mouth pushes the used air out through the opened gill covers.

Air-breathing organs similar to those of labyrinth fishes also occur in pikeheads (*Luciocephalus* sp.) and snakeheads (*Channa* sp.). The question of the relationship of the snakeheads to the anabantoids and the pikeheads has never been settled. Since the snakeheads are designated in some aquarium literature as being closely related to the labyrinth fishes, they are discussed in this book following presentation of the labyrinth fishes.

The dependency of labyrinth fishes on atmospheric oxygen varies greatly among species. Some cannot survive without it, but others rarely utilize it. The African *Ctenopoma* species, for example, can survive for a long time without atmospheric oxygen. Species of the genera *Mulpulutta* and *Parosphromenus* can even get along entirely without supplementary air if they are kept in cool oxygen-rich water.

In 1937, R. Bader, a renowned ichthyologist, carried out comprehensive studies on the development of the labyrinth organ and reached some interesting conclusions. In these experiments he showed that gill breathing alone was sufficient into old age for labyrinth fishes in which contact with the water surface was prohibited from the larval stage. By the use of a specially rigged net, larval fish were prevented

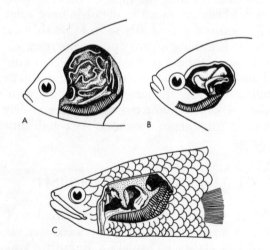

The labyrinth organ of (A) the blue gourami (*Trichogaster trichopterus*) and of (B) the comb tail (*Belontia signata*); (C) the air-intake organ of a snakehead (*Channa* species).

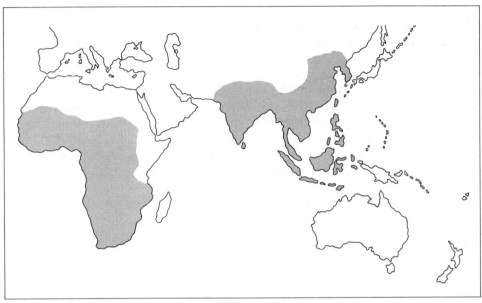

Labyrinth fishes are found in the shaded areas of Africa and Asia.

from reaching the water surface, and a delay in the formation of the labyrinth organ was observed. A control group of larval fish the same age were kept under normal aquarium conditions in which they were allowed to swim to the surface. In this control group the labyrinth organ developed normally. After the net that prevented the first group of fish from reaching the surface was removed, these fish began to develop labyrinth organs. However, development of these organs was completed much later in life than it was in fish of the control group. The evolution of the labyrinth organ can most likely be explained as a well-developed biologic adaptation to waters with a low oxygen content.

The waters inhabited by most labyrinth fishes are rather warm, often reaching temperatures as high as 90°F (32°C). During dry periods the volume of these waters decreases considerably, often resulting in heavy siltation. The siltation combined with a heavy growth of free-floating algae causes these waters to become an opaque "broth." Only fish that can absorb oxygen in

some way other than by the use of gills can survive in such waters. Labyrinth fishes are not, however, found exclusively in such waters. For example, the *Belontia* species of Sri Lanka and the *Ctenopoma* species of Africa occur in somewhat cooler and clearer running water. If one examines the labyrinth organs of various species closely, one finds that there is a correlation between the relative size of the labyrinth organ and the characteristic oxygen content of the water in which the species occurs. In species indigenous to muddy, slow-moving oxygen-poor waters (for example, the *Anabas* species), the labyrinth organ is larger and more extensively convoluted than in species indigenous to highly oxygenated waters.

WHERE LABYRINTH FISH ARE FOUND

Labyrinth fishes occur throughout Southeast Asia and large areas of Africa. In Asia they are found from northeastern China and Korea southward to the Sunda Islands in the Malay Archipelago. To the east they occur in the island groups of the Moluccas and the Philippines, and they are found as far west as Pakistan. Within this tremendous area there are significant climatic differences; therefore the waters inhabited by various labyrinth fishes do not have the same physical and chemical characteristics. Irrigation projects have contributed to the wide dispersion of some labyrinth fishes, so that some species have become predominant forms in artificially irrigated rice paddies. Even though the populations of many labyrinth fishes are densest in standing and slowly moving waters, others occur in swift-running waters—for example, in the higher elevations of Sumatra and Sri Lanka. The common belief that all the labyrinth fishes of Southeast Asia are found in small bodies of turbid oxygen-poor water is an erroneous generalization. These fishes are, in fact, found in almost all types of freshwater habitats, and some species are even found in the brackish water of river estuaries, as well as in some very small water holes and drainage ditches. They occur in irrigation canals on farmlands, as well as in channels and ponds in large cities, as for example in the klongs of Bangkok. In short, there are few bodies of water in Southeast Asia that are not inhabited by labyrinth fishes.

In certain areas of Southeast Asia it is difficult, if not impossible, to determine the original distribution of the residing species. Such species as the giant goramy (not gourami), *Osphronemus goramy,* and the kissing gourami, *Helostoma temminckii,* have been artificially stocked as food fish over wide parts of the area. These fishes are now also found outside of Asia: for example, on the Seychelle Islands, Mauritius and Réunion, in Australia (Victoria), and in South America (Guyana). Several snakehead species valued as food fish have been scattered in a similar distribution. *Channa striata,* for example, which grows to over 3 feet (1 meter) in length, occurs not only in all of Southeast Asia but now also in the Philippines and Hawaii. The extremely hardy Amur snakehead, *Channa argus warpachowskii,* was even introduced into Europe, and snakeheads that have escaped from fish hatcheries have been found in open waters of southern Russia, Romania, and Hungary. Smaller species have also been dispersed into many foreign biotopes. In China and the Malay Archipelago, aquarists have released nonindigenous species of labyrinth fishes into local waters. Many kinds of labyrinth fishes, as well as other kinds of aquarium fishes, such as guppies *(Poecilia reticulata)* and swordtails *(Xiphophorus helleri),* are found in waters that surround aquarium-fish hatcheries.

In Africa, labyrinth fishes inhabit almost the entire southern half of the continent, from Gambia on the west coast eastward along a line through Chad and the Sudan and southward all the way to South Africa. The majority of the African labyrinthid species are found in waters of the rain forests, but some species also occur in other kinds of biotopes. Of the total fish populations of Africa and Southeast Asia, the percentage comprised of labyrinthid species is significantly smaller in Africa. Introduction of nonindigenous species is not as frequent in Africa as in Southeast Asia. During periods of drought, waters outside the rain forests undergo great changes in volume as well as in chemical and physical characteristics. In fact, during the dry season some of these habitats dry up entirely. As expected, species with an accessory air-breathing apparatus have a greater chance of survival than those that lack the apparatus. In African labyrinthids, the correlation between the frequency of air intake and the size of the labyrinth organ is generally not as strong as it is in many of the Southeast Asian species.

HOW LABYRINTH FISH ARE CLASSIFIED

Labyrinth fishes are classified in the order of perchlike fishes *(Perciformes).* Within this order the labyrinth fishes represent one of seventeen suborders. Since labyrinth fishes share many common characteristics, they are often erroneously combined by systematic tabulation into one family, Anabantidae. Most labyrinthid genera have several unique and easily distinguishable characteristics that make placement of individual species in the correct genus a rather easy task. There are, however, exceptions to this, and separation of labyrinth fishes into several new subfamilies would facilitate their identification. So far, no taxonomist has proposed such a system that would be completely satisfactory. According to a classification sometimes seen in old aquarium literature, labyrinth fishes are combined in the family Anabantidae, which is divided into five subfamilies as follows:

1. Anabantinae, which includes the genera *Anabas, Ctenopoma, Sandelia,* and *Helostoma*
2. Macropodinae, which includes the genera *Macropoda* and *Belontia*
3. Osphroneminae, which includes the genera *Osphronemus, Trichogaster,* and *Colisa*
4. Ctenopinae, which includes the genera *Ctenops, Trichopsis,* and *Betta*
5. Sphaerichthyinae, which includes the genera *Sphaerichthys* and *Malpulutta*

Although this classification scheme is occasionally used in aquarium literature, it does not give a true picture of the interrelationships among labyrinth fishes.

In 1963, ichthyologist Karel F. Liem proposed a classification scheme for labyrinth fishes that was an improvement over the older scheme but, based on modern knowledge of these fishes, was still not adequate, for it did not account for behavioral differences among various genera and various species. Liem's classification scheme divided labyrinth fishes into four families, one of which, Belontiidae, was split into two subfamilies. Liem's classification is as follows:

1. Anabantidae, which includes the genera *Anabas, Ctenopoma,* and *Sandelia*
2. Osphronemidae, a monogeneric family that includes only the genus *Osphronemus*

3. Helostomatidae, a monogeneric family that includes only the genus *Helostoma*
4. Belontiidae, which is divided into two subfamilies:
   Subfamily Belontinae, which includes the genera *Belontia, Macropodus, Malpulutta, Ctenops, Trichopsis, Parosphromenus,* and *Betta*
   Subfamily Trichogasterinae, which includes the genera *Trichogaster, Colisa,* and *Sphaerichthys.*

This scheme was based mainly on morphological differences and practically ignored behavioral differences, such as spawning methods. For example, this scheme groups *Macropodus* species, which produce floating eggs, and *Betta* species, which produce sinking eggs, in one subfamily, Belontinae. In the subfamily Trichogasterinae, this scheme groups *Trichogaster* species, which produce floating eggs, with *Sphaerichthys* species, which produce sinking eggs. It has been shown that brood-care behavior varies according to the kind of eggs (floating versus sinking) produced. These distinct differences have not been taken into consideration by Liem or any of the more recent classifiers of anabantoid fishes.

In 1979, ichthyologist Hans Joachim Richter presented a classification scheme for labyrinth fishes that, in the opinion of the author, is well suited for the use of the aquarium enthusiast. This systematic model considers the results of ethological studies as well as morphological information. According to Richter's model, the classification of labyrinth fishes includes four families, one of which, Belontiidae, is subdivided into five subfamilies. Richter's proposed classification is as follows:

1. Anabantidae, which includes the genera *Anabas, Ctenopoma,* and *Sandelia*
2. Osphronemidae, a monogeneric family that includes only the genus *Osphronemus*
3. Helostomatidae, a monogeneric family that contains only the genus *Helostoma*

4. Belontiidae, which is divided into five subfamilies:

   Subfamily Belontiinae, which includes only the genus *Belontia*

   Subfamily Macropodinae, which includes the genera *Macropodus* and *Malpulutta*

   Subfamily Trichogasterinae, which includes the genera *Trichogaster* and *Colisa*

   Subfamily Sphaerichthyinae, which includes only the genus *Sphaerichthys*

   Subfamily Ctenopinae, which includes the genera *Ctenops, Trichopsis, Parosphromenus,* and *Betta.*

In the family Belontiidae, three subfamilies—Belontiinae, Macropodinae, and Trichogasterinae—produce floating eggs and two—Sphaerichthyinae and Ctenopinae—produce sinking eggs.

All labyrinth fishes go back to a common ancestor, a "primeval anabantoid." Fossil material of this fish is known from the lower Tertiary period (about 60 million years ago). Close relatives, older forms not yet equipped with a labyrinth organ, are found in the upper Cretaceous formations (approximately 65 to 100 million years ago). According to some recent research, the chameleon fish *(Badis badis),* or blue perch as it is known in some parts of the world, a popular aquarium fish in the 1930s, 1940s, and 1950s, is said to be a close relative of these ancestral forms.

It has not yet been determined to what extent other families equipped with labyrinth organs are evolutionarily linked to labyrinth fishes. There may be some close evolutionary connection between labyrinth fishes and pikeheads (family Luciocephalidae), which are classified as perchlike fishes of the order Perciformes, as are the labyrinth fishes. Snakeheads (family Channidae), which are thought by some taxonomists to be closely related to labyrinth fishes, are placed in a separate order, Channiformes, in several recent classification schemes. The formation of the labyrinth organ in both labyrinth fishes and snakeheads is assumed to be the result of convergent evolution (that is, a similarity that coincidently came about without the existence of a close evolutionary or phylogenetic relationship). Such convergences are not uncommon among organisms that inhabit a similar type of environment. In other words, it is quite possible that two unrelated species from two different locations could both have developed a labyrinth breathing apparatus. In this regard, it is of interest to note that distribution areas of labyrinth fishes and snakeheads, both in Asia and in Africa, are largely the same. Both have adapted to habitats that temporarily or permanently have oxygen-poor water. Since aside from the auxiliary breathing apparatus there seems to be no close similarities between snakeheads and labyrinth fishes, the presence of the labyrinth organ should not be used as a criterion for establishing a close phylogenetic relationship. Since the question of this relationship has not been subjected to the in-depth systematic study it apparently deserves, it must be considered an open, unresolved, and provisional question and an exciting topic for future study.

APPEARANCE AND STRUCTURE

APPEARANCE

Although labyrinth fishes are classified as perchlike fishes, many of them are more slender or elongate than most perchlike fishes. The bodies of such fishes (that is, *Betta* species or *Trichopsis* species) often show a great litheness or suppleness, whereas those of more stockily built species, such as *Anabas* species or *Ctenopoma* species, seem less sinuous. Most labyrinth fishes are of average length in proportion to body height. In other words, their bodies are about twice as long as they are high and are generally laterally compressed or flattened from side to side, as in the kissing gourami, *Helostoma temminckii*, or most of the *Trichogaster* species. There are, however, species that are more or less cylindrical in shape and much more elongate—for example, species of the genus *Betta*.

Fins

The dorsal fin varies, according to species, in the length of the fin rays as well as in the length of the fin base or the insertion. *Anabas, Helostoma, Colisa, Macropodus, Belontia,* and *Ctenopoma* species, for example, have a very long dorsal fin base that inserts far forward on the back, directly above the pectoral fins in some, and runs all the way along the back, almost to the insertion of the tail fin or caudal fin. On the other hand, species of the genera *Betta, Trichogaster, Trichopsis, Malpulutta, Osphronemus, Parosphromenus,* and *Sphaerichthys* have a short-based pointed dorsal fin, which in some, especially in males and particularly in *Betta* species, is very much like a huge sail. The anal fin is long-based in all labyrinthid species. The number of unbranched hard rays or spines in the dorsal and anal fins varies with species. *Betta* species, for

example, have only one or two, whereas other species, such as *Ctenopoma,* can have as many as fifteen and twenty spines in the dorsal and anal fins, respectively. The soft rays of these fins are bifurcated or trifurcated (multi-branched) in many species.

The pelvic fins are usually located far forward. They are short-based and in some species are located almost at the throat. The pelvic fins have at least one spinous ray followed by a varying number of soft rays, depending on species. In most labyrinthids, the spine of this fin is much longer than the soft rays, and in some species many of the fin rays are more or less grown together into a long feelerlike appendage. The *Trichogaster* and *Colisa* species all have such pelvic fins. These fins not only serve as tactile (touch) organs but equipped with numerous taste buds as organs of taste.

The insertion of the rounded pectoral fins clearly lies below the medial longitudinal axis of the body in all labyrinthid species. As with most freshwater fishes, the pectoral fins have no spinous rays and are quite transparent.

The tail fin or caudal fin of many of the more popular aquarium species—*Trichogaster, Colisa, Macropodus,* and *Helostoma,* for example—is mildly forked and in most has wide, rounded tail lobes. The forked appearance in the tail fin of the *Macropodus* species, however, is deceiving, for the lobes are drawn into long pointed extensions, giving the fin more of a lyre-shaped appearance. The *Sphaerichthys* and *Osphronemus* species also have a very mildly forked tail fin. The caudal fin of *Ctenopoma, Anabas, Sandelia,* and *Belontia* species is rounded or fan-shaped, although in *Belontia* each ray extends a bit beyond the outer edge of the fin, giving it a fringed appearance and thus the common name combtail. In *Trichopsis, Parosphromenus, Malpulutta,* and *Ctenops*

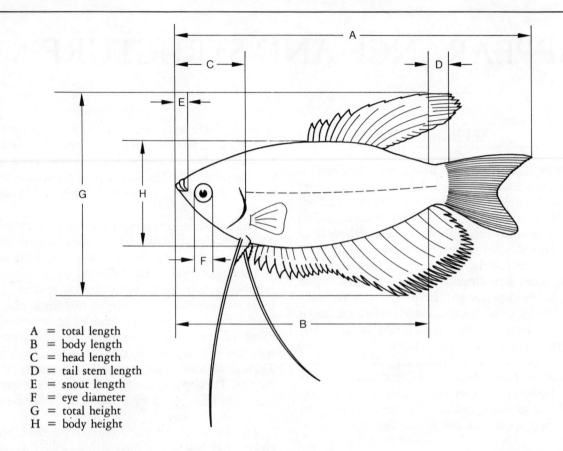

A = total length
B = body length
C = head length
D = tail stem length
E = snout length
F = eye diameter
G = total height
H = body height

species, the caudal fin is rounded, but the centermost rays are extended out to a point, giving the fin a spadelike shape. The caudal fin in the most popular of all aquarium labyrinthids—the domesticated *Betta splendens*—is more or less round in shape, but in males it is very long and flowing, in some strains so long that it hangs downward like a delicate lacy drape.

Scales

The scales of most labyrinth fishes are not especially large, but in some, such as the kissing gourami *(Helostoma temminckii)*, they are exceptionally small for the size of the fish. Squamation consists predominantly of comb (ctenoid) scales. In these, the posterior edge of the scale is serrated, giving the scales a comblike ap-

pearance. Round (cycloid) scales are found, if at all, mostly on the throat, cheeks, and head.

Sex Differences

In most labyrinthid species the sexes can be differentiated by the longer and fuller dorsal, anal, and pelvic fins of the males. In some, the caudal fin is also much larger than that of the female. In most labyrinth fishes there is also a marked color difference between males and females, with the males often much brighter and more colorful than the females. Even though during courtship and mating the color of both sexes intensifies, the color difference between the sexes at that time is much more pronounced than it is during nonmating periods, the males usually being far more brilliant than the females.

Appearance and Structure

INTERNAL ORGANS

Most of the internal organs are located in the visceral cavity, which in most fishes lies in the anterior half of the body. In many labyrinth fishes, however, the viscera are located so far forward that the anal and genital openings lie directly behind the throat, especially in such fishes as the *Colisa* species, all of which have an extremely long anal fin base. In most fishes the anal and genital openings lie just anterior to the anal fin. As in most fishes, the visceral cavity is supported in its entire length by ribs. The upper (dorsal) part of the abdominal or visceral cavity is occupied by a relatively large air bladder. The air bladder serves not only as a hydrostatic organ (for buoyancy) but also as a resonance amplifier for the hearing system, a task often shared by the labyrinth space as well. In their auditory function, therefore, the air bladder and labyrinth organ fulfill the role played by the eardrum and the auditory canal in higher vertebrates. Labyrinth fishes hear well, and many are able to make sounds, but the frequency of these sounds usually lies below the lower limit of human hearing. An exception is the grunting sound produced by the croaking gourami, *Trichopsis vittatus*. Recent experiments have shown that sound production in fishes, including labyrinth fishes, is much more frequent and occurs in more species than had been assumed earlier. The role of sound production and perception in the behavior of these fishes has not been adequately explored.

In addition to the sense of hearing and the combined touch and taste organs of the modified pelvic fins possessed by some labyrinth fishes, the tactile sense of labyrinth fishes is important. By means of this tactile sense these fishes can feel both weak water currents and extremely small temperature differences. The weak water currents are detected by the acousticolateralis, or lateral line system. This is a system of pressure-sensitive cells called neuromasts that are located in pores in the skin. Neurologically, this system is connected to the brain via the eighth cranial nerve, which is the auditory nerve. This system therefore functions as both an auditory and a tactile sensor. In fishes with a visible lateral line, which usually runs horizontally along the sides of the fish, but in a curved arc, the neuromasts are interconnected by a tubelike system. Neuromasts are also located on other parts of the body, especially around the head. Some labyrinth fishes lack a distinct lateral line but have dispersed neuromasts. Tubes connect these scattered neuromasts, but they are not as distinct as in fishes with a well-defined lateral line.

The labyrinth organs are located on both sides of the head, in hollow spaces at the front, back, and above the gill cavities. The most important part of the labyrinth organ is its epidermis, which contains countless fine capillaries. This epidermis is convoluted and branched, which vastly increases the amount of surface exposed to air. Most of the air supply for the labyrinth organ comes through the mouth. Most of the oxygen in the air is absorbed by osmosis through the convoluted epidermis and passes via the capillaries into the bloodstream, where it is circulated throughout the entire body of the fish. The labyrinth organ is completely developed only after a few weeks of life. Therefore, young labyrinth fishes for the first few weeks of life breathe exclusively by means of their gills.

Pikeheads (Luciocephalidae) differ anatomically from labyrinth fishes in a number of ways. They have a deeply cleft mouth, or jaw gape, with protuberant jaws that can swing forward to form a funnellike structure. The anal fin has a deep central notch that almost divides it into two parts, an anterior part and a posterior part. Aside from the first lengthened pelvic fin ray, all fin rays are of the soft-branched type. The air bladder of pikeheads is considerably smaller in proportion to the body size than that of the labyrinth fishes, but nonetheless it functions quite well.

Snakeheads (Channidae) have a long, almost eellike body that is round in cross section. By nature they are as predatory as their large head, deeply cleft mouth, and strong teeth suggest. Their nasal openings are located at the end of extended tubes. Their squamation consists of both cycloid and ctenoid scales. The cycloid scales are particularly large on the upper surface of the head, which contributes to the snakelike appearance of these fishes. The air bladder is large, and it extends almost all the way to the caudal fin base. There are no spines in the anal and dorsal fins. One species, *Channa orientalis,* completely lacks pelvic fins.

HISTORY

IN THE AQUARIUM

The Siamese fighting fish, *Betta splendens,* was kept in aquariums in Thailand long before aquarists of the Western world became acquainted with labyrinth fishes. Male Siamese fighting fish were of particular interest to the Thai because of their natural pugnacious "attitute" toward each other. The Thai used them in staged fights upon which wagers, sometimes of large amounts, were placed. The small containers that housed these fish also served as the arenas for these fights. The Thai selectively bred Siamese fighting fish not for the magnificent finnage and colors for which they are bred today, but rather for their ability to fight, thus enhancing the excitement of the staged fights. In contrast to staged fights between wild Siamese fighting fish, which usually lasted only a few minutes, battles between males bred for their fighting ability often lasted hours. The injuries the fish inflicted upon each other during these fights were mostly limited to torn fins; only rarely did they inflict more serious damage.

One cannot be certain that other species of labyrinth fishes were kept as ornamental pets by the people of Southeast Asia. Because of the ability of most of these fishes to survive the rigors of captivity, however, it cannot be ruled out that the Chinese, who have a long tradition of breeding goldfish, also kept various species of labyrinth fishes in their homes and gardens. Well-known ichthyologists G. A. Boulenger and C. T. Regan reported that the first paradise fish, *Macropodus opercularis,* to reach the Western world, in 1869, were domesticated fish bred by the Chinese. Even though this report cannot be validated, one cannot overlook the possibility that labyrinth fishes were kept in Southeast Asia as ornamental fishes before they reached the West.

Labyrinth fishes were definitely among the first exotic ornamental fishes imported into Europe. That labyrinth fishes survived the strenuous trip at a time when oceanic transportation almost exclusively consisted of relatively slow sailing vessels was no doubt due to their labyrinth organs. The importance of the role of the originally imported labyrinth fishes in the spread of the aquarium hobby cannot really be measured. However, the number of articles dealing with these fishes that appeared in specialized periodicals published before the turn of the century suggests that these fishes were indeed popular and influential in the early development of the hobby. Nearly one fourth of the early literature on ornamental fishes dealt specifically with labyrinth fishes, and the number of species covered could not have been large, because up to the outbreak of World War I only about twenty species of labyrinth fishes were being imported.

By all appearances, the first labyrinth fish species to be imported from the Far East was *Macropodus opercularis,* which reached Germany in 1876; records indicate that these fish were shown at an exhibition in Berlin that year. By systematic breeding selection, Paul Matte, a Berlin aquarist, developed a *Macropodus* strain with particularly long fins. In 1893, he began to market this fish under the name paradise fish. Initially the price for one of these fish was about equal to the monthly salary of a laborer.

Another labyrinth fish, the giant goramy *(Osphronemus goramy),* was bred in Europe by aquarist Carbonnier in 1874. Although these fish, which grow very large, have never become popular as an aquarium fish, young specimens are occasionally imported today.

After several unsuccessful importation attempts, a French aquarist, N. Jeunet, was able to bring the first living Siamese fighting fish into Europe in 1894. Later that year he succeeded in propagating this fish. In his breeding report, published in the *Bulletin Societe d'Aquiculture de France* (1894), sinking eggs were described for the first time in labyrinth fishes. Some of the Siamese fighting fish bred by Jeunet reached Moscow. There, too, breeding was successful. In 1896, Paul Matte saw these fish at a Moscow exhibition. He purchased some of them and took them back to Berlin, where he soon succeeded in breeding them. This species was erroneously described as *Betta pugnax*. The error was discovered when the first true *Betta pugnax* was imported some years later. The fish bred by Matte may well have been the true Siamese fighting fish, *Betta splendens*.

Today most of the labyrinth fishes seen in the aquarium hobby are flown in from Southeast Asia. These are generally not wild-caught fishes but are fishes produced by large-scale commercial breeders in Singapore, Hong Kong, and other Far Eastern cities. As the fish-exporting business began to grow, Oriental breeders developed a variety of color forms in certain species, especially in the Siamese fighting fish, or bettas, as they are more commonly called. In bettas all the color forms seen are true color mutations. This is not true in other species. A number of Oriental breeders of ornamental fishes have developed artificial methods for influencing color in fishes by the use of special foods and hormone supplements.

One can almost count on the fact that labyrinth fishes still unknown in the aquarium trade will be imported in the future. Surprises are certain to come, especially from the Southeast Asia area. Aside from the development of new domesticated strains, the discovery of species still unknown to science is likely, since there are still some habitats that, mainly for political reasons, remain largely unexplored. These areas may eventually open up to exploration. The numerous large and small bodies of water in these areas may hide many a secret of interest to the aquarium hobby.

AS FOOD FISHES

In highly developed Western countries the abundance of fish available for human consumption is limited mostly to those few species caught by commercial methods in oceans or large freshwater lakes. In some less developed countries, however, nearly every kind of fish available, large or small, is used as human food. This is especially true in some inland areas where the means to transport and store ocean fish are lacking. In such areas in Africa and Southeast Asia, only freshwater fishes are known as food fish, among which there are many labyrinth fishes. Their important role as food fish is readily seen in their abundant supply in local markets in these areas, especially in Southeast Asia. In addition to the climbing perch, *Anabas testudineus,* and various snakehead species, small fishes, such as *Colisa* species, are sold as human food. In Indonesia not only are *Trichogaster* species eaten by humans, but even considerably smaller species, such as Siamese fighting fish, end up in the frying pan. Since they have been stocked as food fishes, such species as *Anabas testudineus* and *Channa striatus* are found in the entire Southeast Asia area, including the Philippines. The giant goramy, *Osphronemus goramy,* and the kissing gourami, *Helostoma temminckii,* are also of great importance as food fish. The giant goramy in particular is well-known for its succulent meat, especially in specimens that mature in nutritionally rich brackish water. Under favorable conditions these fish can grow to a length of over 30 inches (70 centimeters) and weigh as much as 20 pounds (about 10 kilograms). The French first released this fish in several tropical countries. There were even attempts to establish it in southern France, but for climatic reasons these attempts failed. After the French found out about this food fish, there were soon no French colonies in which the giant goramy was not established. Today it is occasionally found in Africa and even in South America. Giant goramies and kissing gouramis are bred and raised in large commercial ponds in Southeast Asia.

Most wild labyrinth fishes are caught both with nets and by rod. Climbing perch and snakeheads are dug out of pond bottoms during dry periods, which is when these fishes estivate in the moist mud. Snakeheads, which can grow to over 3 feet (1 meter) in length, have a fine-tasting meat and are almost always sold alive in the marketplaces. Most other labyrinth fishes are also sold alive, because the marketplaces lack refrigeration and dead fish quickly putrefy at high temperatures. If they are protected from direct sun and kept moist, labyrinth breathers survive for several hours out of water. One species of snakehead, *Channa argus warpachowskii,* has become commercially important in moderate climatic zones. Native to the Amur River, it is caught in both Chinese and Soviet territory. In official Soviet fishing statistics the annual catch is given as 35 to 50 tons. This snakehead was also released in other areas of Europe.

TAXONOMY

It is the purpose of systematic models to classify organisms in hierarchically arranged categories. Originally this endeavor was directed mainly at bringing order to an apparently confusing diversity. Only a few easily checked characteristics were used in early classification systems, often leaving to chance whether the chosen characteristics demonstrated structural relationships and phylogenetic connections.

Modern systematic models, on the other hand, take into account as many characteristics as possible, thus attempting to reflect phylogenetic connections. Modern taxonomists are not satisfied with comparing only morphological characteristics but also draw on knowledge from allied disciplines, such as genetics (comparing karyotypes, for example) and ethology (comparing breeding behavior and studying reproductive isolation).

The labyrinth organ allows complete separation of labyrinth fishes from any other suborder of perchlike fishes, and the relatively small number of species makes it easier to survey them for systematic characters than for many other suborders. The species definitions, as determined by eighteenth century taxonomists, coincide astonishingly well with modern findings. Some of the generic characteristics are so precise that individual fish can often be classified in the correct genus by their obvious traits. For this purpose, the identification table of labyrinthid genera (page 23) can be helpful. The individual species can also often be recognized by a simple comparison with a species description. There

are, however, a number of exceptions to this easy identification—for example, in the species-rich genus *Ctenopoma* and in the taxonomically confusing genus *Betta*. In these genera some species have been described more than once, and some characteristics have been interpreted differently by different scientists. Furthermore, these genera are not uniform in their breeding behavior. For example, among them are free spawners (those that merely broadcast eggs freely about), bubblenest builders (those that incubate their young in a nest of mucus-coated bubbles), and mouthbrooders (those that incubate their young in their mouth). The aquarium hobbyist can only hope that the different spawning patterns will be reflected in future systematic revisions of these particular genera.

If the species classification cannot be determined by comparisons with the species description, then fin ray or scale formulas can often help.

Fin formulas are easily determined on most living labyrinth fishes merely by closely observing the fishes when their fins are spread open. Scale formulas, however, are more difficult to determine, even with fishes that have relatively large scales. Scale formulas can be determined best on dead or anesthetized specimens or from close-up photographs, but the most accurate determinations are made on dead specimens. Fin formulas for each species are given in the species descriptions; scale formulas are given only if they are of practical value in the classification.

A = dorsal fin
B = anal fin
C = caudal fin
D = pelvic fin
E = pectoral fin
F = spinous rays
G = soft rays
H = lateral line (the scales from the upper rear
    corner of the gill cover, along the center of the
    body, to the base of the caudal fin)
L = vertical line (unless indicated otherwise, the
    scales of the vertical line are counted downward
    along a line in the greatest depth of the body,
    with the anterior end of the dorsal fin often
    marking the highest point).

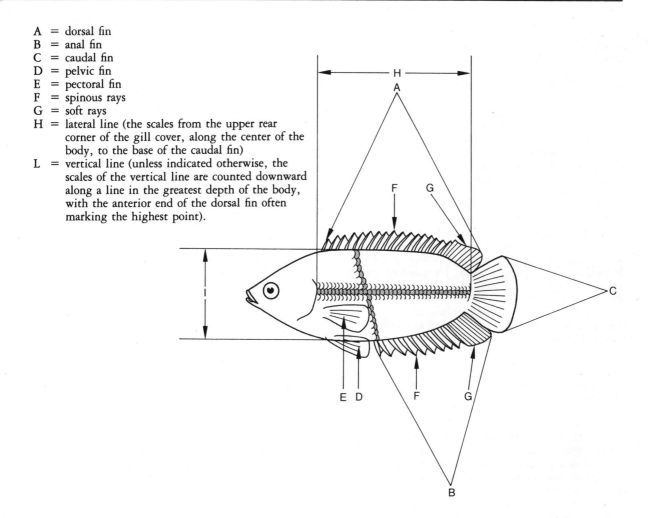

DESCRIPTIONS OF GENERA

Descriptions for determination of genera (details are limited to the most easily recognizable characteristics):

a. *Belontia*
The body is elongate and laterally compressed, with the highest point just behind the head. The dorsal and anal fin bases are almost of equal length. The tips of the spinous rays protrude markedly beyond the soft fin tissue. In older specimens the soft rays of the caudal fin extend somewhat beyond the posterior edge of the fin.

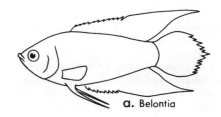

a. Belontia

b. *Macropodus*
The body has an elongate oval shape and is laterally
compressed, with the highest point of the back about
one-third of the way from the head to the tail. The
dorsal and anal fins are long-based and have elongated
rays in the posterior part, which brings the fins to an
exaggerated point, especially in the male. The anal fin
inserts into the body a bit farther forward than the
dorsal fin and tends to be somewhat longer. The
caudal fin is deeply forked in *M. opercularis*, rounded in
M. chinensis. The pelvic fins are long and pointed,
with the first ray much longer than the rest.

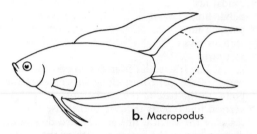
b. Macropodus

c. *Colisa*
The body is oval and highly compressed laterally,
with the high point of the back at about the middle.
The dorsal and anal fins are long-based and almost the
same length, with the spinous rays projecting slightly
beyond the soft tissue of the fins. The posterior end of
the dorsal fin is more pointed than that of the anal
fin, especially in males. The pelvic fins each consist of
one very long fin ray, which when laid back reaches
well past the caudal peduncle. In most species males are
distinctly more colorful than females.

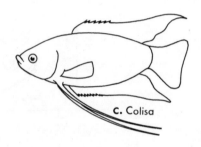
c. Colisa

d. *Trichogaster*
The body is oval but more or less elongate, with the
high point of the back at about the first third of its
length. The anal fin is long-based, with spinous rays
extending beyond the soft part of the fin. The dorsal
fin is short-based and elongated. The outer tip of the
dorsal fin is rounded in females and longer and
pointed in males. The caudal fin is mildly forked.
The pelvic fins consist of a long feelerlike anterior ray
with a few extremely short, barely noticeable rays
attached posteriorly. These pelvic "feelers" reach
backward beyond the caudal peduncle and can be
moved by the fish in almost any direction. There is
very little color difference between males and females.

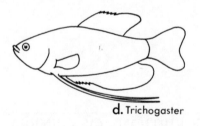
d. Trichogaster

e. *Sphaerichthys*
The body is short and high, with the body height
about half the body length, and is laterally
compressed. The dorsal fin inserts at the high point of
the back, which is about midway along its length.
The fin base is of moderate length and tapers outward
and posteriorly to a point. The dorsal fin is usually
carried in a retracted position. The anal fin is long-
based, inserting into the body ahead of the dorsal fin
insertion, and terminates near the base of the caudal
fin. The posterior tip of the anal fin is rounded. The
anteriormost rays of the pelvic fins are elongate.
The caudal fin is fan-shaped and very slightly forked.

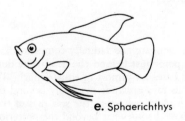
e. Sphaerichthys

f. *Betta*

The anterior part of the body is cylindrical in shape and tapers to a laterally compressed shape in the posterior portion. The pelvic fins are quite long, but unlike others, which have only the anterior rays extended, all the pelvic fin rays are extended, especially in the domesticated *B. splendens*. The dorsal fin is short- to moderate-based and inserts behind the midpoint of the back. It is somewhat elongated in some species, especially in males, but in male domesticated *B. splendens* this fin is exaggerated into a huge sail. The anal fin is long-based, and it inserts into the body far forward, almost under the pectoral fin insertion. The caudal fin is rounded in some species and has a central point in some species. In the domesticated male *B. splendens* the rays of both the caudal fin and the anal fin are very long, so that, when splayed, they billow out like large sails, but when at rest, these fins hang down like long folded drapes.

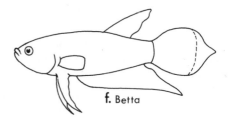

f. Betta

g. *Malpulutta*

The body form is cylindrical, tapering to a laterally compressed shape near the tail. The dorsal fin is short-based and inserts at the center of the back. The posterior rays of this fin in the male are elongated, bringing the fin to a long point. The anterior spinous rays are distinct, rising above the soft fin tissue. The caudal fin is roundish, but the central rays are elongated, giving the fin a distinct central point. The anal fin is long-based, but the rays are short. The pelvic fins are long and pointed.

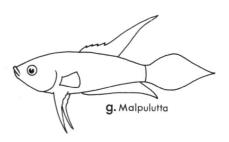

g. Malpulutta

h. *Parosphromenus*

The body is slender and slightly compressed laterally, narrowing in the posterior end. The dorsal and anal fins insert into the body anteriorly at about the same place, but the anal fin extends almost to the tail and the dorsal fin terminates at about two-thirds of the length of the anal fin base. The caudal fin is rounded. The pelvic fins are elongated to a sharp point.

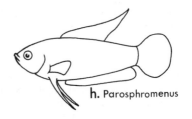

h. Parosphromenus

i. *Pseudosphromenus*

The body is slender and somewhat compressed laterally, narrowing in the caudal peduncle. The dorsal and anal fins are both long-based and of equal length and insert into the body just a bit anterior to the midpoint. The posterior parts of these fins are elongated to a point. The pelvic fins are moderately long. The caudal fin has a roundish diamond shape, and in *P. dayi* a few rays in the center extend beyond the fin.

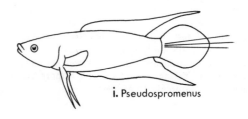

i. Pseudospromenus

j. *Trichopsis*

The body is somewhat slender, laterally compressed, and narrower in the caudal peduncle. The dorsal fin is short-based, inserts into the body at about the highest part of the back (the midpoint), and is high and pointed. The long-based anal fin inserts far forward, almost between the pelvic fins, and the insertion terminates near the posterior end of the caudal peduncle. The posterior end of the anal fin is somewhat elongated and terminates in a point. The pelvic fins are long and pointed and, when folded back, reach almost to the beginning of the caudal peduncle. The caudal fin has a more or less rounded diamond shape and is pointed at the center.

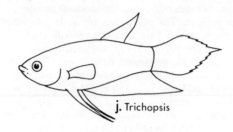
j. Trichopsis

k. *Helostoma*

The body shape is oval and extremely laterally compressed, with a short thick caudal peduncle. The dorsal fin is long-based and inserts just behind the head. It terminates at the caudal fin base. The posterior tip of the fin is rounded. The anal fin inserts behind the dorsal fin insertion and almost between the pelvic fins, just anterior to the midpoint of the body. The anal fin also terminates at the caudal fin base and is rounded on its posterior tip. In both the dorsal and anal fins the spinous rays protrude somewhat, giving the edges of the fins a serrated appearance. The pelvic fins are short and pointed. The caudal fin is fan-shaped, with a very slight central depression, suggesting a very mildly forked shape. The lips are thick and protuberant and are easily turned forward.

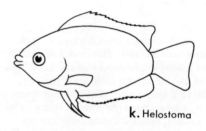
k. Helostoma

l. *Osphronemus*

The body has a long oval-shaped profile and is strongly compressed laterally, with a blunt snout. The dorsal fin inserts at the midpoint of the back and terminates at the beginning of the fan-shaped caudal fin and has a rounded posterior tip. The anal fin inserts behind the midpoint of the body, is short anteriorly, and is elongated to a somewhat pointed posterior tip. The anal fin also inserts posteriorly at the beginning of the caudal fin. The pelvic fins insert far forward and are long and pointed, reaching backward to the midpoint of the anal fin. The lips are thick and protuberant, and the lower jaw is especially prominent. Young specimens have a more pointed snout and unusually large eyes.

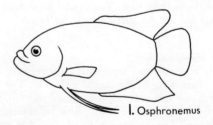
l. Osphronemus

m. *Anabas*

The body is elongate but not slender and is laterally compressed. The dorsal fin has a longer base than the anal fin and inserts into the body just behind the head. The spinous rays of the dorsal and anal fins are prominent, giving the anterior outer edge of the fins a serrated appearance. Both the dorsal and anal fins reach backward almost to the beginning of the caudal fin and are somewhat rounded on the tips. The pelvic fins are short and strong. The caudal fin is fan-shaped with only a slight hint of a central depression on the outer edge. The posterior edge of the gill cover bears two strong spines.

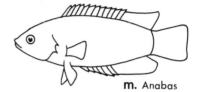

m. Anabas

n. *Ctenopoma*

The body is strongly compressed laterally and is elongate but oval in most species. The exception is *C. ansorgei,* which has a much more slender body shape than most other *Ctenopoma* species. The dorsal fin is long-based, inserting just behind the head and terminating at the beginning of the short, thick caudal peduncle. The rounded anal fin inserts at the horizontal midpoint of the body and terminates directly below the termination of the dorsal fin. The strong spinous rays of the dorsal and anal fins are quite prominent, giving them a serrated appearance on the outer edges. The caudal fin is fan-shaped, with no central depression. The pelvic fins are somewhat rounded. In most species there are two strong spines on the posterior edge of the gill cover. The eyes of these fishes are markedly large.

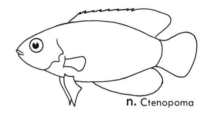

n. Ctenopoma

o. *Sandelia*

The body is slender and laterally compressed. The long-based dorsal fin inserts just ahead of the midpoint of the back and terminates at the thick, short caudal peduncle. The anal fin inserts at about the horizontal midpoint of the dorsal fin, is somewhat pointed, and also terminates at the caudal peduncle. The pelvic fins insert considerably far behind the pectoral fins and are somewhat pointed but not elongate. The gill cover bears two strong spines. The mouth is very deeply cleft.

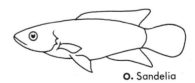

o. Sandelia

SCIENTIFIC NAMES

Animals and plants are given their scientific names according to scientifically established rules. The formation of the name and the use of the different components of the name are subject to internationally binding rules governed by the International Nomenclature Commission. The basic system used in scientific nomenclature was developed by the great Swedish naturalist Carl von Linné and was used by him in the tenth edition of his book *Systema Naturae*. This was the first scientific work in which a uniform nomenclature system was used for the entire animal kingdom (as it was known in 1758). In this binary nomenclature system, each species receives two names. The first is the name of the genus, and the second is that of the species. The generic name is spelled with an initial capital letter, but the species name is not. The names of both the genus and the species are written either in italic type or are underscored. This distinguishing designation is not used for any other higher taxonomic level, such as family or order.

An italicized or underscored name that begins with a lowercase letter and follows the species name designates a subspecies; for example, *Trichogaster trichopterus sumatranus*. Species are divided into subspecies usually only when each subspecies has several clearly distinguishable characteristics of its own and when they are geographically separated from one another.

A complete scientific name includes the name of the individual who described the fish and the year in which the description was published. An example will clarify the system. A blue-color form of *Trichogaster trichopterus* was discovered on the island of Sumatra in 1933. It was described by Ladiges under the species name *sumatranus*. *Trichogaster trichopterus* had, however, already been described by Pallas in 1770 as *Labrus trichopterus*. According to the rules of zoological nomenclature, Pallas is the author of the species. His name and the year in which the species was first described are given after the name *Trichogaster trichopterus* (Pallas, 1770). The author's name enclosed in parentheses means that the described species was subsequently placed in a different genus, that is, one other than that in which it was originally described by the author. In this case it was described originally under the genus name *Labrus* but is currently placed in the genus *Trichopterus*. When a species is currently designated in the genus to which the author first assigned it, the author's name is given without parentheses. By the priority rule of the International Nomenclature Commission, the species name given in the first description is the valid name and may not be changed, although the species may later be placed in a different genus. If the species is later divided into several subspecies, the form on which the species name is based becomes the nominal form. The nominal form can be recognized by the fact that the name of the species and the subspecies is identical, for example, *Trichogaster trichopterus trichopterus* (Pallas, 1770). The newly discovered blue form is correctly designated as *Trichogaster trichopterus sumatranus* Ladiges, 1933.

BEHAVIOR

Much of the everyday behavior of a fish is determined by its food requirements and its enemies. Other aspects of behavior, especially territoriality and aggressiveness are often linked to the reproductive methods of the species.

FEEDING BEHAVIOR

Although most fishes are quite myopic (nearsighted), vision does seem to play a major role in the recognition of food by many labyrinth fishes. In open waters they attack and devour nearly any organism that can fit into their mouths. In addition to live animals, such as insect larvae and small crabs, labyrinth fishes also eat carrion (dead animal matter). Generally they are not particularly choosy in what they eat. In *Trichogaster* and *Colisa* species, the threadlike pelvic fins are equipped with organs of taste, smell, and touch, but these fins seem to play only a minor role in feeding behavior. They are used more for communication between individuals during territorial disputes and mating behavior. These fins could, however, have an important role in finding dead food material, but documented observations are needed to confirm this.

By nature, most labyrinth fishes are carnivores (flesh eaters). Some species, however, also consume plant matter. *Trichogaster* and *Colisa* species eat mostly small organisms, such as insect larvae and daphnia. The kissing gourami, *Helostoma temminckii*, eats these kinds of organisms but also lives on algae and decaying plants. The giant goramy, *Osphronemus goramy*, on the other hand, eats larger organisms, such as shrimp, crabs, and other fish, but also eats aquatic vegetation. In an aquarium these fish can quickly consume all the water plants down to the roots.

Macropodus, Betta, Trichopsis, and *Belontia* species eat small organisms, such as daphnia, but also feed on larger organisms, such as small shrimps, crabs, and worms. This is also true of *Anabas* and *Ctenopoma* species. *Anabas* and *Ctenopoma* species, however, in contrast to their diurnal (active during daylight only) Asiatic relatives, are reported to be crepuscular (active at twilight) and nocturnal (active at night).

To feed on free-swimming food organisms, labyrinth fishes lunge at their prey. Some labyrinth fishes feed on benthic organisms (bottom dwellers) by grazing over the algae-covered rocks and other bottom surfaces, thus ingesting the small creatures that live among the algae fibers. Other labyrinth fishes feed at the water surface by hovering obliquely and more or less sucking in the surface water, thus taking in the food organisms and other items that float on the surface. This form of food intake can be observed in the aquarium when dry fish food is sprinkled on the surface.

A very unusual type of feeding behavior, the spitting down or splashing of terrestrial insects, was first described and photographically documented in a species of *Colisa* by J. Vierke. Archerfishes (*Toxotes* species) are well known for this behavior and have become highly efficient at this type of food acquisition. They shoot a stream of water or a shower of water drops at an insect that is either flying low over the water or has alighted on an overhanging leaf or twig. They are known to shoot with deadly accuracy for a distance of as much as 3 feet (1 meter). *Colisa* species shoot water for a shorter distance. In both *Toxotes* and *Colisa* species, the water is ejected from the mouth by a thrust of the gill covers. The longer range of the archerfish is achieved with a special adaptation of the palate and tongue, which more or less form a tube.

Colisa species do not have this oral adaptation so their spitting behavior is far less effective than that of *Toxotes* species and may be of minor importance in food acquisition.

AVOIDING ENEMIES

Small fishes have many natural enemies both in and above the water. In the water, predatory fishes, snakes, turtles, and frogs lie in wait for them. From above, danger threatens from fishing birds and mammals, including humans. The most effective long-term protection fishes have against all these enemies is their high rate of reproduction. The longer that fishes remain in the larval stage, the greater is the danger from predators. Most fishes are protected from these predators by having a very brief larval stage. In addition to the protection afforded by the high reproductive rate, fishes also possess many different kinds of defense mechanisms.

Hiding is an important defense. Labyrinth fishes living in open water, for example, can flee quickly into the plant thickets or into the bottom mud at almost the very instant a shadow falls across the water, which, of course, indicates the presence of an enemy above (which can be an aquatic or a terrestrial enemy). In hiding, labyrinth fishes remain almost motionless for quite a while. In addition, many species of labyrinth fishes are cryptically colored, which camouflages them when they hide in or near thickets of aquatic vegetation or directly over a coarse rocky bottom. The same can be true of the brightly colored males of some species; that is, the bright stripes or other patterns actually camouflage them by disrupting the predator's view of their apparent body contours. The red and iridescent blue striped dwarf gourami, *Colisa lalia,* is an example of such a fish. Such fishes as *Trichopsis* species are able to remain completely motionless for a long time under the large surface or submerged leaves of certain aquatic plants. Another species that has mastered the art of hiding is *Malpulutta kretseri.* It

is often difficult, even in the aquarium, to detect their presence, as they float completely motionless under submerged plant leaves, making use of every space between the leaves and every curvature of the leaves. It is a coincident curiosity that *Malpulutta* species depend very little on the labyrinth organ for oxygen, and they can therefore remain under water for a very long time, without surfacing for atmospheric oxygen.

School formation is another form of protection against enemies. Dense schools, when seen from a distance by a predatory fish, appear as a large body, which the predator apparently does not readily recognize as food. This is, as mentioned earlier, because most fish are severely myopic. Should the prey be recognized and the school attacked, the fish scatter in all directions, thus distracting the attacker so thoroughly that the attack fails. During the dry season some species of labyrinth fishes (for example, *Colisa* and *Trichogaster* species, as observed by Forselius, Vierke, and others) inhabit the open water areas of rivers, where they form dense schools. In the aquarium, schooling behavior can be observed among the young of *Trichogaster* species. The individual fish within such a school typically move in unison, as though synchronously responding to a single command. For example, they simultaneously swim to the surface for air and simultaneously swim downward again.

TERRESTRIAL MIGRATION

Overland migrations are not uncommon among some fishes. Mudskippers (*Periophthalmus* species), for example, are known for their ability to climb up on emergent tree roots or move about through wetlands. Some armored catfishes, such as the *Clarius* species (one of which is the well-known walking catfish), can migrate overland through wet grasses from one body of water to another. Similar overland migrations are known among some of the labyrinth fishes—for example, in the *Ctenopoma* species.

The best known overland migrator among the labyrinth fishes is the Indian climbing perch, *Anabas testudineus*. (This fish is not a perch, but it is only known by that name.) The German naturalist Marcusa Elieser Bloch (1723–1799) was the first person to describe this species. He received five preserved climbing perch from a missionary active in India, together with the following cover letter: "In India these fish have different names (*Undi-colli, Paunieri,* or *Noza-gri*); these names mean approximately 'tree climbers.' With their saw-like gill covers and sharp fins they climb the palm trees standing on the banks, while rainwater drips down from them." Although later reports on these tree-climbing fish were highly exaggerated, they actually do travel overland. If their habitat dries out, they migrate overland, usually at night when the grasses are wetter, to other nearby water habitats. Evidence has shown that these fish can travel up to 200 yards (180 meters) in a single night. To travel on land the fish rolls its body to the left and right, in short, jerky movements. Forward motion is achieved by the bony pelvic fin rays and the serrated lower posterior edge of the gill covers. Overland migrations may be for these fish the last resort in their fight for survival during the dry season. The chemical composition of the water and the softness of the substrate may be the deciding factors in whether these fish leave their habitat or bury themselves in the moist bottom mud until the next rains come and their habitat is refilled.

Anabas species and *Ctenopoma* species, such as *C. kingsleyae* and *C. multispinnis,* are quite capable of surviving embedded in moist bottom mud. During the rainy season, when the water is high, these species migrate from the rivers to the vast adjacent floodplains, which is where they reproduce. When the habitats formed by the flooding rivers begin to recede, these fishes are frequently caught in small ponds and waterholes. This triggers the fishes' overland migrations toward the rivers. Although overland migrations are part of the normal behavior of some labyrinth fishes, most of them never leave the water.

TERRITORIALITY

Most aspects of territorial and fighting behavior in labyrinth fishes are related to reproduction. Usually only during breeding periods, while caring for their broods, do male labyrinth fishes form territories. These territories are zones in which no other individuals of the same species, except gravid (ripe) females, are tolerated. Other species are also chased out of the territory, but the tolerance toward other species is often greater than it is toward males and nongravid females of their own kind. Males defending their territory attack even larger fishes. In a confrontation with a member of the same species the intruder almost always loses the challenge and is often injured, even if it is considerably larger than the territory-defending male. This is probably because the territory defender may be more strongly motivated.

In general, in mouthbrooding species (species that incubate their eggs and young in their mouths), territorial behavior is not as apparent as it is in the nest-building species.

How much space the males of bubblenest-building species claim for their territory differs from species to species and also depends largely on environmental conditions, such as the temperature, the chemical condition of the water, the amount of water movement (currents), the amount of light, and the kind and amount of vegetation. In dense beds of vegetation, the numerous hiding places provided by the plants allows the territories to be quite small. In water that is relatively free of plants, individual territories are considerably larger. In a reasonably large aquarium, several males can stake out territories adjacent to one another. For example, in a 50-gallon aquarium containing dense beds of vegetation, four or five males can establish territories in which nests will be built. If, on the other hand, the vegetation is sparse or absent, perhaps only two males will build nests and they will usually be as far away from each other as possible. If in a densely planted aquarium several males are unable to establish territories, they can still move around in the aquarium rather freely; they are attacked by the territory-guarding males only when they become directly visible to them, and then they have the opportunity to hide between the plants. If there is no vegetation, however, the males that were unsuccessful in building nests have no place to hide and are attacked. With the more aggressive species these wanderers are often killed.

When an aquarium is overcrowded with labyrinth fishes, territorial formation usually does not occur, especially if the aquarium is sparsely planted. Under these circumstances, it is not uncommon for many labyrinth fish species to form schools. This corresponds to their behavior in the wild during the nonbreeding season.

AGGRESSIVENESS

In many labyrinth fishes, aggressive or antagonistic behavior is prevalent at some stage of the life cycle. In most species this aggression begins in the middle to late juvenile stage. In some species, such as the giant goramy, *Osphronemus goramy,* this behavior wanes as the fish mature. In others, such as the snakeskin gourami *(Trichogaster pectoralis),* the kissing gourami *(Helostoma temminckii),* or the combtail *(Belontia signata),* the aggressiveness seen in juveniles remains in adulthood. There are, however, some labyrinth fish species, such as the Siamese fighting fish, *Betta splendens,* in which the degree of aggressiveness in males, especially toward their own kind, increases dramatically as the fish mature. There are also some labyrinth fish species that never develop much aggressiveness, except at breeding time. Some examples of the latter are the dwarf gourami, *Colisa lalia,* and the honey gourami, *Colisa chuna.*

Although some labyrinth fishes are by nature more aggressive than others, even among closely related species, it should be noted that much of the aggressiveness seen in any aquarium fish is an artifact resulting from poor aquarium conditions, such as improper temperature, improper water chemistry, improper diet, inadequate shelter areas (plant thickets), or overcrowding. In a well-planned aquarium or in nature, where there is no crowding and an abundance of vegetative cover, males participating in territorial disputes are rarely killed. Injuries do occur, however, particularly in the aquarium and especially to the intruder. Therefore, if more than one male of a strongly territorial species is to be kept in an aquarium, no matter how well planned the aquarium, the fish should be watched very carefully. The fighting itself may not kill the fish, but infectious diseases resulting from injuries often do. If you are breeding the more aggressive territorial labyrinth fishes, culling the weaker fish as the brood matures will help reduce injuries. Even if you are not breeding your labyrinth fishes, nonbreeding males should be removed from the aquarium once the dominant males have begun nest building.

Threatening Postures

When it comes to a dispute between two males of approximately equal dominance, the two opponents first "try" to intimidate each other by making themselves appear as large as possible. This is accomplished by splaying out the fins as far as possible. In addition, in some species, such as the Siamese fighting fish, *Betta splendens,* the gill covers are turned outward and the dark red branchiostegal membranes are extended out of the gill cavity. When viewed from the front

A rice paddy about 6 miles east of Bangkok, Thailand. Rice paddies are the habitat of numerous labyrinth fish species almost everywhere in Southeast Asia. When the rainy season begins, the fishes invade the warm waters of the flooded rice fields in order to reproduce.

by another fish with low visual acuity (as is the case with most fishes), the displaying fish appears much larger than it really is. An actual attack rarely follows this threatening posture. The opponents then assume a side-by-side position, either head-to-head or head-to-tail. Their bodies, with the fins still splayed, are then laterally twisted into an S shape, and the fish beat currents of water against each other or even slap one another with their tails, the sides of their bodies, or their heads. This kind of display rarely results in injury, but it gives the inferior fish (most often the intruder) the opportunity to submit and retreat. The display may be repeated several times if the intruder persists.

Fighting Rituals

If a dispute continues, the fish circle each other and attempt to bite each other on the tail. If this does not result in retreat of the intruder, then in some fishes, such as *Belontia, Macropodus, Trichogaster,* or *Betta* species, a mouth-to-mouth battle may ensue. In these battles the fish attack each other with their mouths wide open, and they often lock jaws and begin to wrestle, pushing and twisting each other, sometimes rather violently. Such fishes as *Trichogaster* or *Betta* species, which depend heavily on atmospheric oxygen, usually interrupt the battle to go up to the surface for a bubble of air. Such fishes as *Belontia* species, which are much less dependent on atmospheric oxygen, may continue the battle with their jaws firmly locked onto each other for a very long time, as long, perhaps, as an hour or more. No matter what behavioral phases the battle proceeds through, one fish, usually the intruder, capitulates.

TOP: A natural klong near Bangkok. These watercourses are the natural habitats of many kinds of fishes, including labyrinth fishes. Here fishing is done with a casting net. Labyrinth fishes are an important element in the diet of the Thai.
BOTTOM: In the larger waterways of Thailand, labyrinth fishes as well as snakeheads are caught with large sinker nets.

The capitulating fish usually indicates its submissiveness by clamping its fins close to its body and assuming an oblique, head-up position. The fish then promptly retreats, usually to a position beyond the boundary of the defender's territory. The intruding fish is rarely pursued beyond that territorial boundary. The escaping intruder does not often sustain any serious injuries, but in certain more aggressive fishes, such as *Belontia* and *Macropodus* species, serious injuries can sometimes occur, especially in a small overcrowded aquarium in which there are few adequate hiding places. Injuries are inflicted by the fish ramming and biting each other, and in the more aggressive species, death of the submissive fish is not uncommon, particularly in the aquarium.

In many labyrinth fishes, the individual phases of the fighting rituals are not clearly distinguishable. For example, certain parts of the fighting behavior may be omitted without the fish having to go back through the whole ritualized cycle. Frequently, in *Colisa* species, the individual phases of the ritual follow each other with such rapidity that it is difficult to tell when one phase stops and the next begins.

In *Trichopsis* species, grunting sounds are made that may be an early part of the fighting ritual. In croaking gouramis *(Trichopsis vittatus),* for example, submissive males do not even participate in the fighting rituals if they are properly "croaked" at by a more dominant male. If a battle ensues, the croaking sounds seem to be omitted.

When a male labyrinth fish first begins to establish a territory, a female venturing into the area may be attacked. This frequently occurs in *Colisa, Macropodus,* and certain *Betta* species, all of which are strong bubblenest builders. There may be a correlation between the strength of the bubblenest itself, the strength of the male's territorial defense of the nest, and the ferocity with which the intruding female is attacked. This behavior could also be the result of mistaken identity, since in some species the color pattern taken on by a submissive male closely resembles the mating color pattern of a female. Furthermore, this behavior occasionally occurs in species that are not so strongly territorial.

A peculiarity that sometimes occurs, even in species that are strongly territorial, is that the male may allow the female to participate in the early defense of the territory, in bubblenest construction, and even in brood care. This rare behavior is infrequently seen in *Macropodus* and *Pseudosphromenus* species and is hardly ever seen in any other species. In other species this may only be an artifact produced by the conditions of captivity.

The ritualized fighting behavior in labyrinth fishes occurs, as a general rule, only during initial formation of the territory. If another male of the same species appears within the territorial boundaries once mating has occurred and brood care is underway, the ritualized displays and tail slapping are omitted and the intruding fish is immediately attacked and chased off, as would be a fish of any other species.

BREEDING BEHAVIOR

Despite the fact that there are many species-specific differences in mating and brood-care behavior in labyrinth fishes, all these fishes can be categorized into two main breeding types: those that produce floating eggs and those that produce sinking eggs. Moreover, among the species that produce floating eggs, there are some that build floating nests of mucus-coated bubbles and brood their eggs and young in these nests and others that do not. Among the species that produce sinking eggs, there are some that build elaborate bubblenests in which to brood their eggs and young, and others that incubate their eggs and young in the buccal cavity (mouth).

Floating eggs contain small oil droplets that render them lighter than water; hence they float. During the course of embryonic development, the oil collects in the yolk sac as a single drop-

let. Since the yolk sac and the oil droplet are on the ventral side of the embryo, the embryo develops belly-up. At a later stage of development the oil becomes more dispersed around the yolk, and this allows the newly hatched embryo to assume a belly-down position.

Sinking eggs lack the oil droplet and are therefore heavier than water. In bubblenest-building species that lay sinking eggs, the eggs are usually gathered by the male in his mouth as they descend through the water. After the male has picked up a mouthful of eggs, he "spits" them up into the floating bubblenest, where they are secured between tightly packed bubbles. The hatched larval fish adhere to the bubbles by means of sticky mucus organs located on the head.

Brood care in bubblenest builders that lay sinking eggs often demands the expenditure of more effort than that required of species that lay floating eggs. This is because floating eggs can develop even without a bubblenest while they float freely on the surface. Sinking eggs, however, will not usually hatch if they remain on the bottom. On those rare occasions when such eggs do hatch, the larvae may not survive if left to themselves, for as new hatchlings they do not have the power necessary to propel themselves to the surface and hold themselves there, which is the only place where they can survive the first crucial week or so of life. The brooding male constantly picks up the eggs and larvae that drop out of the nest and spits them back into the nest.

Brood care in mouthbrooding species requires even more effort than that expended by any of the bubblenest builders. In these species, the eggs, which sink to the bottom, are picked up in the mouth of the male, and there they are kept until they hatch a few days later. The male continues to keep the larval fish in his mouth until they absorb all of the yolk. In *Betta brederi*, for example, which is a typical mouthbrooder, the male releases free-swimming young about a week to ten days after the spawning. During the incubation period the fish spends most of its time at the surface breathing moist atmospheric air. Little if any food is consumed during the incubation period.

Number of Eggs and Survival Chances

The brood-care behavior of species that lay sinking eggs is more specialized than that of the species that produce floating eggs and probably developed later in the evolutionary history of these fishes. Also, within each of the two groups, the simplest forms of brood-care behavior probably led the way to more highly developed forms. Parallel with increasing specialization in brood-care development, the number of eggs produced in a spawning seems to have decreased. For example, in those species that produce floating eggs and exhibit little or no brood care, the number of eggs produced is very high; well over 2000 in most cases and as high as 5000 in some. In species that produce floating eggs, build loosely constructed bubblenests, and provide a minimal amount of brood care, the number of eggs is more typically near 1000. In species that produce floating eggs, build well-organized bubblenests, and provide complete brood care, the number of eggs produced is typically well under 1000 and as low as 150 to 200 in some species. Brood-care behavior in all species that lay sinking eggs is more highly developed than in any species that produces floating eggs. The least specialized of these, *Betta splendens,* can lay as many as 500 eggs. Species that build more complex nests, using clumps of plants and bubbles, rarely lay more than about 150 eggs. Finally, mouthbrooding species, the most specialized of all forms, do not often lay more than about 50 eggs, and often the number is much lower.

Paralleling the reduction in the number of eggs produced is an increase in the percentage of the brood that survives the early stages of development. Survival of the species is thus ensured in those fishes that produce a small number of eggs. A large number of eggs is the only way to ensure survival of the species in those fishes that provide little or no brood care.

Different Levels of Care

If one views the various forms of brood care provided by labyrinth fishes as a series of overlapping developments rather than a step-by-step progression of one form to the other, the overall brood-care behavior in these fishes can be more easily understood. For example, among species that produce floating eggs there are those that, after depositing the eggs, no longer care for them and often even eat them. Such behavior is commonly found in kissing gouramis, *Helostoma temminckii,* and in some species of the genus *Ctenopoma.* Among these species there may be a few free-floating clusters of bubbles produced but nothing even resembling a nest. There are, however, species that guard eggs floating freely on the water surface. This type of brood care is typically found in *Belontia* species. It also occurs occasionally in species that normally build bubblenests but in which nest building is sometimes omitted. Well-organized bubblenests consisting only of layers of mucus-coated air bubbles are built by *Macropodus* species and by most *Trichogaster* and *Colisa* species. The dwarf gourami, *Colisa lalia,* and the moonlight gourami, *Trichogaster microlepis,* build very sturdy bubblenests that are reinforced with pieces of plant matter. The giant goramy, *Osphronemus goramy,* builds nests of various types, both at and under the surface of the water, depending on prevailing environmental conditions.

Among species that lay sinking eggs, the members of the South African genus *Sandelia* are the only ones that merely deposit eggs near the bottom, among the plant leaves, and guard them there; *Sandelia* species do not build a nest at all. Except in members of this genus, this form of brood-care behavior is rarely seen in labyrinth fishes. There are, however, reports that this form of brood care occurs in *Pseudosphromenus* species, which normally build bubblenests, if they live under stressful environmental conditions. The deposition of eggs on the roofs of caves with both male and female participating in the placement of the eggs and with no bubblenest construction has been observed and photographically documented in *Parosphromenus deissneri*. This species adds some mucus-coated bubbles to the nest site after spawning is complete, apparently to support the eggs on the roof of the cave. Other *Parosphromenus* species, according to a personal communication from the well-known aquarist Walter Foersch, build bubblenests in the cave before spawning begins. Some species that lay sinking eggs build bubblenests just under the surface, under large floating leaves. Others build their bubblenests below the water surface inside hollow spaces, such as between plants, or under large horizontally oriented leaves. In *Trichopsis* species, the eggs are deposited in hollow spaces in egg packets, each containing several eggs. The females in some of these egg-concealing species sometimes actively participate in brood care.

Finally, there are those labyrinth fishes that lay sinking eggs that are incubated in the mouth of the male, although the female in some of these species will initially pick up the eggs and carry them to the male. Most mouthbrooding labyrinth fishes are members of the genus *Betta*, but of course, not all members of this genus are mouthbrooders. In the chocolate gourami, *Sphaerichthys osphromenoides*, it is the female that incubates the eggs and young in the mouth. This form of spawning has been described in aquarists' literature and has been photographically documented. The very early stages of bubblenest construction also occur in this species. Several times the author has observed the release of a shower of fine bubbles as a prelude to courtship and also observed that eggs were spit into this foam. These eggs, however, were subsequently picked up again by the female and orally incubated.

The Bubblenest

The primary function of a bubblenest is to confine the eggs to one place so they may be cared for properly by the attending parent fish. In the case of floating eggs, the nest prevents the eggs from drifting away. In the case of sinking eggs, the nest helps keep the eggs afloat. To prevent the nest itself from floating away, it is often anchored to the floating parts of firmly rooted plants. After the eggs hatch, the larval fish anchor themselves to the bubbles in the nest by means of sticky head glands.

Bubblenest construction can be quite different in each species and can even vary among different individuals of a single species. The presence or absence of floating plants can influence the size and type of nest an individual fish builds. Although the integral use of plant material in bubblenests occurs in only a few species, there are species that prefer to build their bubblenests under clumps of floating plants. This causes the floating plants to be lifted only slightly. A bubblenest so constructed is not easily recognized from above by predators; thus the fish guarding the nest is protected.

Water surface tension also influences the size and type of bubblenest. Water with a fairly high mineral content has relatively strong surface

Spawning behavior of *Macropodus opercularis*, a species typical of those that lay floating eggs. (I) The male (left) splays its fins and swims back and forth in front of the female in an effort to lure her under the bubblenest. (II) The female (right) swims to the bubblenest. (III) The female swims up to the side of the male, which is now bent into a U-shape. (IV) In the embrace the pair turns in such a way that the female is now upside down. (V) The eggs and sperm are ejected into the water, and the eggs rise toward the bubblenest as the pair separates from the embrace.

The later stages of mating in *Betta splendens*, a typical bubblenest builder that lays sinking eggs. (VI) The released eggs sink across the splayed anal fin of the male. (VII) First to be free of the embrace, the male gathers the sinking eggs in his mouth and will spit them up into the bubblenest.

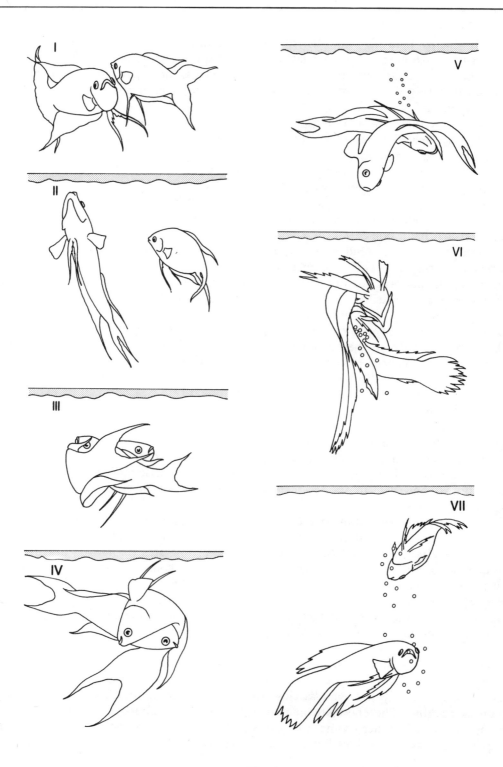

tension. This helps hold the bubblenest together, making it stronger and longer lasting, and enables it to be larger and denser than a nest built in relatively mineral-free water with lower surface tension.

In some species there are variations in nest construction that may be related to different local environmental conditions, such as water chemistry or temperature. For example, it has been reported that the giant goramy, *Osphronemus goramy*, which lays floating eggs, in some waters of India builds a submerged, hollow, spherical nest that has a side entrance. The nest is built with plant matter, mud particles, and mucus-coated air bubbles. This construction is unusual for a fish that lays floating eggs. Submerged bubblenests are usually seen only in species that lay sinking eggs. It is unknown, however, whether this type of nest construction is the rule or the exception in the giant goramy, for in other areas it has been reported that this species builds a floating bubblenest containing pieces of plant matter. Only floating nests have been reported from breeders in the Western world. It is not unusual, in a species with a distribution this wide, where environments vary markedly from one location to another, that different forms of nest building occur. It seems to be quite normal, however, in species that lay sinking eggs and have a limited distribution, that the nest is built either submerged or floating, depending on local conditions at the time. For example, *Malpulutta kretseri, Betta imbellis,* and most *Pseudosphromenus* species are known to build their nests either way in the same locality but at different times.

Environmentally influenced deviations of brood-care behavior are not rare in wild fishes. Such deviations are not observed as often in the aquarium, however, because aquarium environments are generally maintained by most good aquarists in a very steady state and as close to optimum as possible. The chocolate gourami, which, as mentioned earlier, sometimes spits the eggs into a mass of bubbles before finally beginning oral incubation, is a widely distributed species found on the island of Sumatra and also across large parts of the Malay Peninsula. The living conditions of these fish vary considerably over this broad geographic area, and there are within this range a number of relatively small isolated populations. These are called island populations. (Island population is a name given to any population that is geographically isolated from the main population, even in the middle of a large land mass. Such an island population could be isolated, for example, by altitude, as a result of a tectonic earth movement.) Island populations of the chocolate gourami have been isolated from the main populations for a long time, and each has evolved in somewhat different ways. Thus among some imported fish rudiments of an older type of brood-care behavior remain. Assuming, in the course of evolution, that mouthbrooding developed from the care of sinking eggs in a bubblenest, the presence of intermediate stages between these two extremes of brood-care behavior can be explained.

As mentioned earlier, bubblenest-building males do not tolerate other fish in close proximity to the nest, especially during its construction, and then even females of the same species are usually chased away. There are variations in this behavior in species that produce floating eggs. For example, males of *Trichogaster* and *Colisa* species chase away approaching females of the same species, whereas *Macropodus opercularis,* as a rule, tolerates the female near the nest, even during its construction. The behavior of a male labyrinth fish toward a female is not always dependent on the presence of a nest. If, for instance, a ripe female makes her presence known to the male by pushing against his side with her mouth, then spawning can follow in some species without the presence of a bubblenest. This behavioral variation is known to occur in the blue gourami, *Trichogaster trichopterus,* and in the giant gourami (not goramy), *Colisa fasciata.* The missing nest is built later by the male either between matings or after all spawning activity has ceased. In species that lay floating eggs it is not unusual to see females swim from bubblenest to bubblenest, spawning with several males in succession. Seen even more frequently is the spawning of one male with two females in quick succession. Spawning without specific pair formation can be observed only in very large aquariums but probably is not uncommon in nature, at least in *Colisa* and *Trichogaster* species. The changing of partners during spawning does not seem to occur in *Macropodus opercularis,* which lays floating eggs, or in any bubblenest-building species that lay sinking eggs. True pairing, though temporary, with mutual participation in brood care occurs in *Belontia* species.

Courtship and Mating

Although many varieties of brood care are seen in labyrinth fishes, courtship behavior is quite similar in all of them. A courting male displays to a ripe female by almost instantaneously developing intensified colors and splaying out his fins in an apparent attempt to lure the female under the bubblenest or to a selected spawning site. The ripe female shows her readiness to spawn by quickly swimming up to the male and nudging him in the side with her mouth. These signaling rituals occur in mouthbrooders and bubblenest builders as well as in free-spawning species, such as the kissing gourami, *Helostoma temminckii.*

Following a few such displays and nudging responses, the first of several pseudomatings occurs. The female swims up to the side of the male, and the male begins to arch his body around the female. At this point in the embrace, there is a departure between bubblenest builders (and free spawners) and mouthbrooders in the position of the mating pair. In bubblenest builders the female turns upside down as the male arches himself in a tight upside-down U shape over the female's belly, so that both his head and tail are pointing downward, and the genital openings of both fish come into close apposition. In mouthbrooders, both fish more or less arch or cup themselves around each other in such a way that their genital openings are brought together, but the female does not turn upside down. In some of their embraces the male may arch himself into a U shape under the female, with his head and tail pointing upward.

At the culmination of the embrace, in both bubblenest builders and mouthbrooders, both fish quiver several times and then the embrace is released. Usually by the second or third repetition of the nudging and embrace the pseudomatings become real matings, as eggs and sperm are showered into the water, upward in bubblenest builders and downward in mouthbrooders. These various types of embraces ensure fertilization of almost all the eggs. In bubblenest builders, including those that lay sinking eggs, the embrace and egg release usually occurs high in the water, close to the bubblenest. In mouthbrooders the embrace usually occurs in midwater or closer to the bottom substrate.

Care of the Eggs

In species that lay floating eggs, the eggs slowly rise to the surface under the bubblenest (if there is one). Those eggs that reach the surface but miss the nest are gathered in the mouth of the male and spit up into the nest. In bubblenest builders that lay sinking eggs, the laterally splayed anal fin of the male catches many of the eggs, thus slowing their descent through the water. This facilitates the male in gathering most of the eggs before they reach the bottom substrate, and then he spits them up into the nest. He also retrieves those eggs that have fallen to the bottom and places them in the nest. During egg gathering, the *Betta splendens* male chases the female away from the nest, permitting her back only after he has gathered all the eggs and is ready for another embrace. In the pointed-tail paradise fish, *Macropodus cupanus,* on the other hand, it is the female that does most of the egg gathering. (Unlike its close cousin *M. opercu-*

laris, which produces floating eggs, this bub-blenest builder lays sinking eggs.)

In mouthbrooders, usually the male gathers the eggs in his mouth, except in the case of the chocolate gourami, *Sphaerichthys osphromenoides.* In this species, as mentioned earlier, the female orally incubates the eggs.

In all cases, the spawning embrace is repeated over and over again, as often as twenty-five times in larger females, until the female's ripe egg supply is exhausted.

In all bubblenest builders the nest is guarded usually by the male alone after spawning is concluded. Occasionally in some species the male allows the female to share the brooding duties, but this is a rarity. Like other rare behavioral events that have been described in this book so far, this, too, may be an artifact of captive life. The attending parent continuously repairs the nest, adding bubbles as they are needed to fill holes that develop through drying out or through damage caused by intruding fishes or defending activities. Should any of the sinking-type eggs fall from the nest, the attending parent picks them up and spits them back into the nest. Newly hatched larvae are kept in the nest in the same manner as the eggs.

As soon as the young fish have absorbed all of the yolk and begun to swim freely, which is usually on the third to the fifth day after hatch-ing, brood care ceases. The same is true for mouthbrooding species. In these species, the egg-carrying parent remains just under the water surface, usually in dense clumps of vegetation, until the free-swimming larval fish are released from the mouth. Unlike the larvae of mouth-brooding cichlids, once the young of mouth-brooding labyrinth fishes are released, they are not taken back into the mouth.

REPRODUCTIVE HABITS OF RELATED SPECIES

The mating and brood-care behavior of snakeheads (Channidae) is similar to that of some labyrinth fishes. Mating takes place close to the surface. The male arches around the female and turns her upside down. The eggs float to the surface, where they are kept together by the male until the larval fish are free-swimming. According to some recent observations, mouth-brooding care is offered in one snakehead species.

Little is known about the reproductive habits of the pikeheads (Luciocephalidae). There are, however, some indications that they are mouth-brooders, but successful reproduction in captivity is so rare that the details are not yet clear.

CARE IN AN AQUARIUM

IN A COMMUNITY AQUARIUM

Many labyrinth fishes can be kept together in community aquariums with other species. For example, some of the *Trichogaster* species, such as the pearl gourami *(T. leeri)*, all of the *Trichopsis* species, and most of the *Colisa* species, even the larger ones, can be kept together with very small fishes, if the aquarium is spacious and well planted. Small fishes would be in danger only when the labyrinth fishes are building bubblenests and spawning, and then generally only when they swim into the nesting territories. Small fishes must have enough room to escape from territory-defending labyrinth fishes, and they must have plenty of hiding places.

The Siamese fighting fish *(Betta splendens)*, known simply as the betta, the males of which are extremely aggressive toward their own kind, can be kept safely in a community aquarium with smaller fishes if only one male is kept in the tank. However, a significant danger of keeping even one male betta in a community aquarium is not so much to the other fishes but to the betta itself, for many other kinds of fishes tend to nip at the long, flowing, brightly colored fins of the male betta. The male betta's tankmates should be selected very carefully. Many of the barbs, such as the tiger barb *(Capoeta tetrazona)*, and the characoids (tetras), such as the black tetra *(Gymnocorymbus ternetzi)* which are known fin nippers, should be avoided. Female bettas are more peaceful toward each other and toward other fishes. Furthermore, their shorter, more drably colored fins do not provoke attacks from other species. Therefore, any number of them can be kept together in a community aquarium.

Young kissing gouramis *(Helostoma temminckii)*, one of the largest species of labyrinth fishes, behave rather peacefully toward smaller fishes. As they mature, however, they become progressively more aggressive toward each other and toward smaller fishes. *Macropodus* species, young or old, can be quite aggressive toward other fishes, especially smaller ones. Even the popular albino *M. opercularis,* which is known to be more docile than the normally colored strain, should be watched carefully in the presence of smaller fishes.

Juvenile giant goramies *(Osphronemus goramy)* are often imported for the aquarium trade. These fishes grow quite large, outgrowing average-size aquariums in almost no time. In large aquariums, however, they can be kept quite successfully with some of the larger Central and South American cichlids, such as *Aequidens rivulatus* or *Astronotus ocellatus.*

There are some genera of labyrinth fishes in which most members are simply too aggressive to be kept with smaller fishes. For example, most *Belontia* and *Ctenopoma* species will attack and devour smaller fishes. With enough space and plant shelter, however, these fishes get along well with each other and with other territorial fishes, such as cichlids. These fishes can be successfully kept in the aquarium together with other fishes that in nature are found in the same habitats. For example, in nature *Belontia signata,* commonly known as the combtail, is found together with the Asiatic cichlid species *Etroplus maculatus.* In an aquarium containing fast-moving aggressive species, the latter tends to be a bit shy, but they are strongly territorial. They seem to fare well with *Belontia signata,* as long as adequate shelter (plant cover, rocks, and so on) is provided. *Ctenopoma* species share their riverine habitats with some African cichlid species and some catfishes. In the aquarium they all get along well together if plenty of refuge is provided and the aquarium is sparsely populated.

Small, shy, slow-moving labyrinth fishes, *Parosphronemus* species, *Malpulutta kretseri*, *Trichopsis* species, and *Sphaerichthys osphromenoides* (the chocolate gourami), should be kept only with other small fishes having a similar temperament. These species tend to hide and not eat well in busy community aquariums.

Community aquariums populated exclusively with labyrinth fishes can be very attractive. Such an aquarium should have a capacity of about 20 gallons (18 liters), but a larger tank would be even better. The water should be maintained at about 78°F (23°C) with a thermostatically controlled aquarium heater. The aquarium should be well lit for about twelve hours a day, by either natural sunlight or artificial illumination, or a combination of both. Illumination of the aquarium promotes healthy plant growth and the establishment of suitable microflora and microfauna, which helps maintain the chemical balance of the water.

Any aquarium containing labyrinth fishes should be tightly covered, as most of these fishes are agile jumpers.

The aquarium water should be well filtered, not only to keep the aquarium clean but also to promote a uniform distribution of heat. Using a filter to promote water circulation eliminates the need for supplemental aeration with an airstone. Since, in nature, most labyrinth fishes typically live in slow-flowing swamps, streams, and rivers, large power filters, which provide a very strong water current, are not needed, except in very large aquariums.

Since only a few species of labyrinth fishes require special chemical conditions in the water, no special filtering material is needed for most of them. For those few species that require soft acidic water, this condition can be provided by filtering the water through a bed of sphagnum peat moss. This is done by sandwiching a layer

of clean, boiled peat between two layers of standard polyester filter floss. For most labyrinth fishes, however, the standard filter setup, which consists of a layer of carbon sandwiched between two layers of polyester filter floss, is quite adequate.

Plenty of plants of both the rooted and floating types should be used. Aquarium plants are not only decorative but also provide excellent hiding places for smaller fishes or fishes with a shy temperament. Some labyrinth fishes, such as *Trichopsis* species, like to hide under an overhanging leaf but in a more or less open area. Floating plant leaves provide such a refuge. Other labyrinth fishes prefer to hide in the nooks and crannies formed between the stalks and leaves of submerged plants. Although not always available because of their seasonality, the tiger lotus (*Nymphaea lotus*) and the yellow water lily (*Nuphar luteum*) meet both requirements. In moderate light they grow large fanlike submerged leaves; in brighter light these fanlike leaves reach the top of the aquarium and float horizontally on the surface. Less seasonal and thus more readily available are the *Cryptocoryne* species, which grow well in moderate light and are less demanding in their care. They produce submerged lanceolate leaves that grow to a variety of heights, depending on species, and fan out, thus providing the kind of shelter preferred by most labyrinth fishes. Extremely hardy, easy to care for, and quite commonly available is the underwater banana plant (*Nymphoides aquatica*), which in moderate light produces round fanlike submerged leaves that closely resemble those of the yellow water lily. In strong light this plant sends long stringy stalks to the surface where the fanlike leaves flatten out and float like small lily pads. Many other suitable varieties of plants are commercially available. Aquarium literature that describes them in detail may be found.

For stocking a community aquarium with labyrinth fishes, there are a plethora of different species combinations from which to choose. Generally, greater success will be achieved by combining bubblenest-building species that are not too closely related than by combining those that are. This is because very closely related species tend to be more aggressive toward each other than species more distantly related. With careful selection and good care, most of the labyrinth fishes selected will readily build bubblenests and spawn. Not suitable for a community aquarium are highly predatory fishes, such as *Belontia* species and many of the *Ctenopoma* species. Also unsuitable for a community aquarium are such fishes as the chocolate gourami *(Sphaerichthys osphromenoides)*, *Malpulutta kretseri,* and other similar species, all of which require special water conditions and special care.

IN THE SINGLE-SPECIES AQUARIUM

A single-species aquarium, as the name implies, is one in which fish of only one species are kept. This type of aquarium is advantageous for keeping those species that have special requirements of space, water chemistry, temperature, and food. A single-species aquarium also lends itself well to those situations in which special observations or studies are to be made.

Malpulutta, Parosphromenus, and *Sphaerichthys* species thrive better in soft, mineral-poor water that is slightly acidic. If the available tap water is soft (that is, it has a total hardness not exceeding 6 DH, it should be filtered through peat moss until its pH value stabilizes at about 6.0. If only hard tap water is available, it will be necessary to mix it with demineralized water that has been filtered through peat moss for about one week. The two kinds of water should be mixed together in whatever proportion is necessary to reduce the hardness to about 5 or 6 DH. If the tap water has a hardness of 20 DH, for example, it should be mixed in a 1:3 ratio of hard water to peat-filtered water. The total hardness of the water should not be allowed to

drop below about 3 DH; otherwise the pH value, in this case the acidity, will fluctuate rapidly and erratically.

WATER CHANGES

It is believed by some aquarists that because labyrinth fishes can breath atmospheric air they are less sensitive than other fishes to water pollution. There is no basis for this erroneous belief. An overaccumulation of ammonia in the water, for example, causes the same skin, gill, and organ damage in labyrinth fishes that it causes in any other aquarium fish. The difference is, however, that because labyrinth fishes can breath atmospheric air, they can cope with these irritants a little longer than other fishes. To reduce the concentration of these irritants, which arise from fish wastes and other kinds of decaying organic matter, aquarium water should be partially replaced at regular intervals. Although the filter may remove most of the free-floating particulate matter and even some of the detrimental ions that accumulate as by-products of organic decomposition, chemicals, such as ammonia, nitrites, nitrates, carbonates, and others, remain in the water and must be removed by other means. Replacement of some of the water dilutes these pollutants to less harmful concentrations. The frequency of water changes needed depends basically on the density of the fish population of the aquarium, but it also depends somewhat on other factors, such as the feeding routine and water temperature. In an aquarium with a relatively dense population, about one-third of the water should be replaced with fresh water of the same chemical composition and temperature about every two to three weeks. If, however, only a few small fish are kept in a larger, well-planted aquarium, water changes may be necessary only every two to three months. In an aquarium containing a large brood of young growing fish, on the other hand, it is particularly important that the water be partially changed as frequently as possible, for in such an aquarium feeding is usually quite heavy and more wastes than normal are being produced. In this situation a partial water change should be made at least once a week.

NUTRITION

FEEDING ADULT FISHES

In nature, labyrinth fishes are found mostly in slow-moving or standing nutrient-rich waters, where aquatic insects and their larvae, flying insects that fall onto or alight on the water surface, worms, small crustaceans, and larval fishes are for most species the main items in the diet. Some species of *Anabas* and *Ctenopoma,* as well as snakeheads and pikeheads, prey on other juvenile and adult fishes. Some labyrinth fishes are by nature more vegetarian, and algae are the main plants eaten by most of them. Numerous microorganisms live in mats of algae found in open waters and are consumed together with the algae. The giant goramy *(Osphromenus goramy)* eats some of the higher forms of aquatic vegetation. This fish is known as the "water pig" in its native land and is a true omnivore that rejects almost no food.

The shape and size of the mouth allow some predictions about the natural food of a fish. For example, shallow clefts in the jaws of *Helostoma, Trichogaster, Colisa,* and *Sphaerichthys* species suggest that these fishes eat mostly small food organisms. The deep jaw gape of *Macropodus, Belontia,* and *Betta* species, as one would predict, allows them to eat larger organisms, including some nonlarval fishes. The protrusable jaws of *Anabas* and *Ctenopoma* species suggests that these fishes are also predators that feed on nonlarval fishes.

Proteins, Carbohydrates, and Fats

All these food organisms provide labyrinth fishes with the essential nutrients they need for daily survival. The most important of these nutrients, protein, contains all the amino acids animals need to build cells. They must consume this protein in the form of other animals, because animals, unlike plants, cannot synthesize their own amino acids from raw minerals. Lack of protein leads fishes and other animals to a variety of deficiency symptoms that can lead to death. Carbohydrates and fats (also essential elements of the diet) can be stored in the body of an animal, but protein cannot be stored in any appreciable amount and hence must be supplied to the animal almost continuously. Insects and insect larvae as well as worms and other fishes are among the richest sources of protein available to fishes.

Carbohydrates and fats also play a major role in the diet of fishes. They are the primary sources of energy and make possible all of the body functions, including the almost continuous process of protein synthesis from metabolized amino acids. An excess of fats and carbohydrates, however, can be responsible for many disorders in labyrinth fishes, including severe digestive upsets. Excesses of these basic food elements are stored as deposited fat. Too much deposited fat stresses these fishes and reduces their reproductive capabilities, and in some cases even causes sterility. Particularly life threatening in labyrinth fishes are fat deposits in and around the liver. Only a few studies have been done on how much fat and carbohydrate a fish can assimilate without harm. These studies have shown, however, that there are considerable differences among species in their ability to absorb fats and carbohydrates. These studies have also shown that growing juveniles can absorb considerably more fat and carbohydrate than can adult fishes and that these juveniles can convert most excesses into growth without the formation of fat deposits. Because of the particular problem labyrinth fishes have with fatty livers, excess amounts of fatty foods should be avoided.

Because of the lack of information on this subject, it is difficult to say just how much fat represents an excess amount. Experience has shown, however, that fatty livers can be avoided in labyrinth fishes if the amount of fat in the diet is kept below 8% of the total diet.

Vitamins, Minerals, and Trace Elements

Part of a balanced diet is an adequate supply of vitamins, minerals, and trace elements. Generally, the aquarium hobbyist need not worry too much about these vital food components. They are present in sufficient quantity in live foods and in most high-quality, fresh dry food preparations. It is important that dry foods not be stored too long, especially in containers that have been opened. The efficacy of the vitamin content of dry foods decreases rapidly when the foods are stored too long or stored improperly. Once the containers are opened, they should be closed tightly after each use.

Roughage

Roughage is another important substance in the diet. Although it has little, if any, nutritional value, its presence in the diet is crucial for proper digestion. Roughage in the labyrinth fish diet is provided by the chitinous exoskeleton of insects and insect larvae as well as by other invertebrates and the plant fiber (cellulose) of algae and other aquatic vegetation. Both cellulose and chitin should be in the diet of labyrinth fishes. In the case of those fishes that are predominantly insectivorous, many of the organisms they eat feed on algae and other vegetation, so these fishes consume cellulose indirectly in the gut content of their prey. Digestion is by no means only a physiological process but is also a mechanical process. Chitin and cellulose in the stomach help grind food mechanically to prepare it for absorption in the intestine. Without adequate roughage, digestive disorders and malnutrition can occur.

Live Food

Living organisms seem to be the best food for labyrinth fishes. Individuals raised wholly or partly on living foods appear to fare much better than those raised on commercially prepared dry foods alone. On live foods they grow faster and larger, are usually more colorful, are more likely to spawn, produce a higher number of eggs, offer better brood care, are less likely to cannibalize their own brood, and have greater disease resistance. Why this is so is not really known. There are many theories, including the idea that movement of the food evokes a more vigorous feeding response, thus they eat more. Because most fishes are extremely myopic, it is unlikely that vision plays much of a role in the feeding response. Most prey organisms produce substances called pheromones. Using their keen olfactory sense, fishes may detect these pheromones. Most fishes also have keen senses of

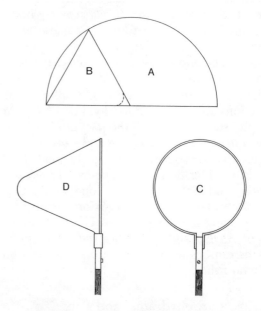

Pattern for cutting a net for the catching of infusoria (dust food). Part A folded and sewn together (or glued with fabric glue) results in net B. The lower end is rounded off.
C = Net frame
D = the net seen from the side

hearing and touch, via the swim bladder, the weberian ossicles, the otoliths, and the lateral line, or acousticolateralis system. This enables them to detect the most minute vibrations in the water, which may evoke a strong feeding response. Live food organisms apparently provide more stimulation to all of the fishes' sense organs than do dry foods, and therefore it is likely that the fishes will consume more food on a live food or partial live food diet than they otherwise would.

Suitable live foods are mosquito larvae, bloodworms (midge larvae, *Chironomus* sp.), glassworms (phantom midge larvae, *Chaoboris* sp.), water fleas (*Daphnia* sp.), brine shrimp (*Artemia* sp.), tubificid worms (*Tubifex* sp.), white worms (*Enchytraeus* sp.), and chopped earthworms. Of these, mosquito larvae may be the best all-around food in terms of nutritional value, but they are not that easy to come by. They must be scooped from standing waters that usually team with all sorts of organisms, many of which can harm fishes. These waters are often fetid and sometimes inaccessible. Bloodworms are found in the mucky bottoms of the same waters. These creatures can kill small fishes directly by eating through their stomachs with their sharp mandibles, before the fishes' digestive enzymes kill them. Since tubifex worms live around sewage dump sites, they carry with them a whole host of pathogenic bacteria that can cause disease in the fishes that eat the worms. Daphnia have a very high percentage of chitin and a proportionately low percentage of protein. Therefore, as a major item in the diet, daphnia can cause digestive disorders in the fishes that eat them. White worms, or enchytraeids, on the other hand, are unusually high in fat, so using them as a major part of the diet can also bring on health problems in fishes. Nutritionally speaking, brine shrimp may be as good an all-around live food as mosquito larvae, and they are much more readily available. Dry brine shrimp eggs are packaged in cans and are easily hatched. Without too much difficulty, baby brine shrimp can be raised to maturity in just a few weeks. Adult brine shrimp are about ¼-in. (6 mm) long and are suitable as food for most adult labyrinth fishes. Live adult brine shrimp can be purchased in many pet shops in and around large metropolitan centers. Like any other food, they, too, can cause digestive upsets and deficiency diseases if they are fed to the fishes too often and in too large a quantity.

If large quantities of these food organisms are collected at one time, they can be stored for subsequent feeding almost indefinitely by freezing them. They retain almost all their nutritional value and can be conveniently used at will. Most of the live food organisms mentioned here are also flash frozen and packaged commercially. Frozen fish foods, including chopped beef heart, chopped clams, and other meaty substances, are available in many tropical fish stores.

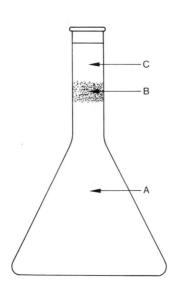

Schematic description of the growing of pure cultures of *Paramecia*.
A = Culture liquid
B = cotton plug
C = stale fresh water

Freeze-drying is one of the latest methods of preparing fish foods commercially. It is possible to purchase freeze-dried mosquito larvae, tubifex worms, brine shrimp, daphnia, and other small crustaceans. Although there is little documented evidence that freeze-dried foods are as nutritious as live foods, labyrinth fishes seem to eat them greedily and seem to thrive fairly well on them.

Other Suitable Food

A variety of commercially prepared dry flake foods are available in tropical fish stores. Most of these foods have a high nutritional value but often, unfortunately, have too high a percentage of carbohydrates. Analysis of the ingredients and their corresponding nutritional values are usually printed on the packages. Most dry foods are made predominantly from meaty substances, such as shrimp meal, fish meal, and beef meal. However, flake foods made predominantly from vegetable matter are available for fishes, such as kissing gouramis (*Helostoma temminckii*), which require more vegetable matter in the diet than most other labyrinth fishes. If needed, a good substitute for green flake foods is canned or frozen spinach. Also, lettuce leaves placed in water and exposed for several days to direct sunlight quickly become covered with green algae. After a few days, these partially decomposed lettuce leaves can be given to the fishes. Giant goramies (*Osphronemus goramy*) also eat lettuce leaves.

Another good alternate food is frozen meat, such as beef or fish. Small pieces of meat can be frozen and then scraped, allowing the scrapings to fall into the aquarium water. The meat, whether it is beef or fish, should be as lean as possible, for some fishes seem to develop digestive problems from a direct intake of fat, especially beef fat.

FEEDING YOUNG FISHES

If young labyrinth fishes are to be raised to full-sized, healthy breeders, they must have the right kind and amount of food right from the start. The fry of some labyrinth fishes, such as the *Betta* species, are so small when they first start to swim and feed that they can consume only microscopic food organisms or dust-fine powdered dry foods. Some of the young of the more aggressive species tend to grow much faster than many of their siblings because they get more than their share of the food. This situation can be avoided or at least improved by providing the entire brood with an abundance of food, but care must be exercised to clean up any leftover food before it causes the water to become foul.

Microscopic Live Food

The use of live foods for fry and rapidly growing juvenile fishes is perhaps more important than it is for adult fishes. Prepared dry foods are not taken as readily by these young fishes as live foods, nor in as great a quantity. However, the collection of microscopic live food organisms is not an easy task. Specialized equipment, such as very fine mesh nets of different sizes and mesh gauges, is required to strain and sort the organisms. Also, daily trips to collecting sites are necessary, unless special holding facilities are built to keep a supply of these microorganisms alive. Furthermore, suitable pollution-free collecting sites are not very accessible to most urban or even suburban dwellers. In collecting these microorganisms there is always the danger of unknowingly collecting some microscopic creatures, such as hydra, which are capable of

TOP AND BOTTOM RIGHT: Aquarium fishes are sold in the Sunday Market in Bangkok. Here, short-finned wild-type Siamese fighting fish bring very high prices. They are purchased for use in staged fights on which some very high wagers are made. In the upper photo, in the right-hand jar, two male Siamese fighters are engaged in such a fight. BOTTOM LEFT: In the same market, climbing perch (*Anabas*) species are sold for human consumption. Snakeheads are in the containers covered with netting; if they are kept moist, they can survive long journeys.

devouring the fry instead of being devoured. To offset this danger, the microorganisms must be separated from any larger organisms that may have been collected by pouring the collection through a series of nets with a mesh size large enough to let the desirable microorganisms pass through and trap the potential enemies.

An easy and safe way to ensure having an ample supply of live food organisms for young labyrinth fishes is to cultivate them. For the smallest fry, single-celled protozoans, such as *Paramecium caudatum,* or small but simple multicelled organisms, such as rotifers, make excellent first foods. These and many other such microorganisms are easily cultured in small aquariums or wide-mouthed 1-gallon mayonnaise jars.

There are many different ways to prepare a culture medium. One of the simplest is to boil a handful of hay and leaf litter for about fifteen minutes in water from a stagnant pond or from an old aquarium. This accelerates the breakdown of organic matter that will later serve as food for the microorganisms. When the solution cools to room temperature, it is poured, along with the hay and leaf litter, into the culture container. Then some fresh unboiled water from the stagnant pond or the old aquarium is added to the container; thus the living microorganisms in the old water are inoculated into the culture. If pond water is used as the inoculum, it should be inspected for fish enemies, such as hydra or

TOP: A top view of the bubblenest of a paradise fish (*Macropodus opercularis*). The reddish yellow eggs are easily recognized between the bubbles. The color of the eggs is not peculiar to the species but is caused by coloring matter in the food.
BOTTOM LEFT: The black paradise fish (*Macropodus concolor*) has long puzzled aquarists, for its origin is still unknown, even though it has been kept as an aquarium fish since the late 1930s.
BOTTOM RIGHT: The Celonese paradise fish (*Belontia signata*), also known as the combtail, has a misleading name since it is not very closely related to the genus *Macropodus*.

predatory insect larvae, before it is poured into the culture container. This is easily done by pouring it into a white porcelainized pan where any potential fish enemies may be easily seen and removed. An even easier way to start a culture is to use any of several commercially made preparations as the inoculum. These are usually sold in the form of tablets that, when dropped in the culture water, trigger the growth and reproduction of any microorganisms that are in the culture. In using these tablets, it is not necessary to collect pond water and boil it. The tablets can merely be dropped into the aquarium water—in some cases right in the aquarium with the new hatchlings. Each manufacturer includes specific instructions with the tablets. The cultures are usually kept separately from the aquarium in which the fry are living. After a few days of storage (uncovered) in a relatively cool dark place, the culture will be teaming with a variety of microorganisms that are suitable as food for the smallest labyrinth fish fry. The fry should be fed frequently (four or five times a day) so that they have an almost continuous supply of food. A small light placed close to the culture container will attract the microorganisms to that location. They can then be easily removed from the culture with a kitchen baster.

Larger Food

After about four or five days of feeding on microorganisms, the fry can be switched to larger foods, such as newly hatched brine shrimp (the nauplii of *Artemia salina*) or microworms (tiny nematode worms cultured as fish food). These organisms can also be cultured quite easily. Microworms are raised in small dishes of pabulum kept at room temperature. The worms congregate on the surface of the pabulum, where they can easily be scraped off with a Popsicle stick or a single-edged razor blade. The scraping is then swirled into the water containing the fry.

Dried brine shrimp eggs can be purchased in any tropical fish store. Cultured water for brine shrimp is prepared by mixing ordinary table salt with dechlorinated tap water. Since brine shrimp come from a number of different localities, it is best to use the proportions of salt and water listed on the container. Wide-mouthed jars with 1-quart or 1-gallon capacity are used for the cultures. The culture water should be heavily aerated. About a fourth of a teaspoon of dried brine shrimp eggs is added to each quart of water. At room temperature the eggs will hatch in about forty-eight hours. This process can be accelerated to twenty-four hours by keeping the water at about 80°F (26.5°C). Once the eggs have hatched, the brine shrimp nauplii can be collected, as with the microorganisms, by attracting them with a small lamp placed close to the jar. The live shrimp are seen near the light as a large reddish orange vibrating mass. They can be removed from the culture using a kitchen baster and should be washed free of salt before feeding them to the fish. This is done by placing them in a fine-mesh net (sometimes sold as a brine shrimp net) under gently running tap water, of about the same temperature as the water from which they were taken, for a few seconds. The net is then everted into the aquarium. The baby brine shrimp will live only about an hour in fresh water. It is therefore best to feed the fry only the amount that can be consumed in a few minutes. When their bellies become markedly swollen and take on the orangish hue of the brine shrimp contained therein, the fry have had enough. Better growth will result if the fry are given numerous small feedings each day rather than fewer large feedings. This is true of any food they are given.

Microworms and brine shrimp nauplii are good first foods for larger fry, such as those produced by *Belontia signata*.

BREEDING IN THE AQUARIUM

Some labyrinth fishes, such as the chocolate gourami, *Sphaerichthys osphromenoides,* are rather difficult to spawn in the aquarium, but most labyrinth fishes can be induced quite easily into spawning in captivity. In terms of the number of young produced and saved for rearing, the best results are usually achieved by breeding a species in its own aquarium, rather than in a community aquarium containing an assortment of other fishes. This is especially true of mouthbrooding and bubblenest-building species that lay sinking eggs, for parental brood care is usually required until the young begin to swim free. Were breeding to occur in the community aquarium, few of the young would escape the hungry mouths of the other residents of the tank. On the other hand, the eggs and young of species that lay floating eggs, even those that build bubblenests, do not require much brood care. Therefore, if they spawn in the community aquarium, the eggs can be scooped from the surface in a net and placed in another small prepared aquarium for hatching. The fry would subsequently be moved to larger quarters for rearing. In the case of those fishes that have built a bubblenest, the nest, eggs and all, can be scooped out with a shallow saucer for placement in a hatching tank. No matter where the fishes are bred, you should not try to raise more of the brood than you have tank space to accommodate. This is especially important with highly fecund species, such as the *Trichogaster* species, or the kissing gourami, *Helostoma temminckii,* which may produce thousands of eggs in a single spawning.

SPECIES THAT LAY FLOATING EGGS

Most species that lay floating eggs are very fecund. Therefore, if you wish to raise large broods, you must have large aquariums to accommodate them. The aquarium must, of course, provide the optimal environmental conditions. To ensure successful breeding you must also watch the fish carefully, and intervene if necessary.

The Right Aquarium

As said earlier, labyrinth species that lay floating eggs are very fecund. In bubblenest-building species, the number of eggs in one spawning can be anywhere from 400 or 500 to 2000 or 3000 or more. Even more productive are the free spawners, such as the kissing gourami, *Helostoma temminckii,* and some of the *Ctenopoma* species, which produce only a scattering of bubbles and lay as many as 4000 to 5000 eggs or more in a single spawning. You can breed a pair of blue gouramis, *Trichogaster trichopterus,* for example, in a 5-gallon aquarium, but it is not possible to raise the 800 to 2000 fry they will produce in such a small tank.

Large Enough: Aquariums of 20 gallons or more are necessary for good results in raising a brood of *Macropodus opercularis* or the smaller *Colisa* species, such as the dwarf gourami, *Colisa lalia.* Aquariums of 30 gallons or more are recommended for large *Colisa* species, such as the giant gourami, *Colisa fasciata,* and for most *Trichogaster* species. For the most fecund species of *Trichogaster,* such as the blue gourami, *Trichogaster trichopterus,* or the snakeskin gourami, *Trichogaster pectoralis,* a 50-gallon aquarium is necessary, if you expect to raise a majority of the brood. Equally large aquariums are required for breeding and raising the fry of the kissing gourami, *Helostoma temminckii,* large *Ctenopoma* species, and similar large fishes. Aquariums of 125 gallons or larger are necessary for breeding the largest of the labyrinth fishes, the giant goramy, *Osphronemus goramy,* and the snakeheads (*Channa* species).

Heavily Planted: The best breeding results are usually achieved in older aquariums that have become well balanced both chemically and biologically. The aquarium should be heavily planted with rooted and floating plants. Not only do the plants help the aquarium come into chemical and biological balance, but they provide sheltered areas, which most males prefer for breeding, and they also provide nesting materials, which some species use in bubblenest construction. The plants also provide an abundance of hiding places for females, most of which are treated rather roughly by males during courtship and spawning.

Many kinds of plants can aid the environmental balance of the aquarium and at the same time provide for the specific needs of the fishes. A plant especially well suited for use with labyrinth fishes is the floating liverwort *Riccia fluitans.* This nonflowering plant grows in dense, spongy masses at the water surface and, under certain conditions, may also grow in a submersed form. The spongy clumps of this plant, which is commonly called riccia, harbor numerous microorganisms that can serve as food for newly hatched fry. Its soft leaves are easily broken apart by most fishes; thus small pieces of riccia are used by some bubblenest-building fishes, such as the *Colisa* species, to reinforce their bubblenests. Others build their nests in, under, and around such masses of floating vege-

tation. To some labyrinth fishes, such as kissing gouramis (*Helostoma temminckii*) and giant goramis (*Osphronemus goramy*), riccia can be a dietary source of vegetable matter. Under moderately strong light and in clean, clear water, riccia grows very rapidly and can completely overrun the aquarium unless it is pruned almost daily.

Water sprite (*Ceratopterus thalicroides*) is another plant that is suitable for labyrinth fishes. It grows either floating or submersed, under the same environmental conditions as riccia. In the floating form, dense tangles of roots hang downward and provide the same kinds of shelter and food sources as riccia. Some labyrinth fishes, such as the *Trichopsis* species, are rather shy and do not thrive well in stronger illumination; thus riccia and water sprite would not do well in an aquarium properly illuminated for these species. However, another plant that grows in dense clumps but does well in dim light is the Java fern, *Vesicularia dubyana.* This plant attaches itself to rocks and driftwood and grows in dense masses around them. It makes an especially good haven for harassed female labyrinth fishes.

Correct Temperature and Chemistry: While some bubblenest-building labyrinth fishes that produce floating eggs thrive well in coolish water (such as the blue gourami, *Trichogaster trichopterus,* and the paradise fish, *Macropodus opercularis*), most require fairly warm water (78 to 82°F; 23 to 24°C) and will not spawn in water that is any cooler. Specific breeding temperatures for each species are given in the species descriptions.

Most labyrinth fishes are not very demanding insofar as water chemistry is concerned. They thrive and breed in water that is mildly alkaline (pH 7.6) to mildly acidic (pH 6.6) and in water that is moderately hard (20 to 30 DH) to moderately soft (3 to 6 DH). Just make sure that extremes of either parameter are avoided and that the fishes are not exposed to any sudden and drastic changes of pH, DH, or temperature.

Courtship and Spawning

With most bubblenest-building species, whether the eggs are the floating or the sinking type, the males tend to harass the females during courtship and spawning, almost brutally in some cases. Therefore, in small aquariums it may be necessary to remove the female as soon as spawning is complete. The male does not usually wander far from the nest; therefore the female is usually safe in a larger aquarium, for she can swim beyond his range.

Courtship in species with sinking eggs or floating eggs is about the same. In both this behavior is highly ritualized, with minor differences among species. At the right temperature and water conditions for the species, the male responds to the presence of a gravid female by splaying out his fins and twisting his body into a sigmoid shape. This is usually carried out in the area selected by the fish as a spawning site. This display behavior may be repeated many times, apparently in an effort to draw the female into the spawning territory. During this courtship period the male often pursues the female and even attacks her. Shortly after the courtship begins, if the fish is a nest-building species, the male begins to construct a nest comprised of mucus-coated bubbles and, in some cases, small pieces of plant matter. Once the nest is complete, the male intensely pursues the female until she finally swims under the nest or to the area the male has designated as the spawning site. There, a number of embraces occur, in various positions, depending on species, with each embrace resulting in a shower (usually in an upward direction) of eggs and sperm. After each embrace the female usually retreats from the territory as the fertilized eggs rise to the surface. The male gathers in his mouth the eggs that have, on rising, missed the nest or the spawning area, and he spits them into the nest or into the general area in which the other eggs have risen. Another courtship display occurs, and the female returns to the spawning site, where another embrace ensues. This sequence of behaviors continues until the female is devoid of ripe eggs. In most species the female is then chased away from the nest and the male remains at the site, where he tends the nest and keeps the eggs from floating away.

Care of the Young

For the next two to five days, depending on species and water temperature, the young remain in the nest, where they absorb the remainder of the yolk sac. Once the yolk sac is completely absorbed, the fry begin to swim free and feed on their own. At that time the attending parent or, in a few cases, parents should be removed from the aquarium, as they will now begin to cannibalize the young. In the case of the free spawners, those that do not build a formally structured bubblenest or build no semblance of a nest at all, the parent fish should be removed from the aquarium as soon as spawning is complete, for these species offer little or no brood care and are likely to begin devouring the eggs.

SPECIES THAT LAY SINKING EGGS

Since most of the species that lay sinking eggs are relatively small fishes and lay far fewer eggs than those species that lay floating eggs, they can be successfully bred in small aquariums. Of all these species, only the Siamese fighting fish (*Betta splendens*) produces a relatively large number of eggs (as many as 500 in one spawning). Because of the expected brood size and the fact that the males usually inflict considerable damage on the females, an aquarium of at least 10 gallons capacity or larger should be used for breeding this species. A 5-gallon aquarium can be used for all other species that lay sinking eggs.

It should be remembered in setting up aquariums for species that lay sinking eggs that some of these species prefer building their bubblenests submerged, under a shelter of some sort. For this purpose, suitable overhanging leaves or other appropriate items must be available. The leaves of the larger *Cryptocoryne* species, a flat piece of cork floating on the surface, half a floating Styrofoam cup, a cave under rocks or a piece of driftwood, or even a broken clay flowerpot laid on its side can all serve as suitable nesting sites for these fishes. The caves formed by rocks or broken flowerpots are also suitable as spawning sites for *Parosphromenus* species, which prefer to deposit their eggs under a firm supporting surface.

Warm temperatures aside, there are no special water requirements with respect to pH or hardness in the breeding tank of mouthbrooders. They are more likely to reproduce, however, in an aquarium that is well planted and contains plenty of other sheltered areas as well.

RAISING THE YOUNG

The fry of labyrinth fishes must be well cared for. They must be provided with the type of food they require and with the proper aquarium conditions. They must also be culled to allow uniform growth rate and in some species to prevent injuries during aggressive behavior.

Food Supply

The demise of the fry of most aquarium-bred labyrinth fishes is the result not of disease but of starvation. With an adequate food supply even a large brood can be reared successfully. In the species descriptions the best first food organisms for each species is discussed. In all species, feeding should not be started until all the fry are swimming free. With extremely small fry it is often difficult to tell if the food, which itself is barely visible to the naked human eye, has been eaten. The best way to tell is to examine the fry

with a small magnifying lens. If they have been eating, their tiny bellies will be round and probably pinkish.

It is difficult to see if all the small food organisms have been eaten, for they often die and fall to the bottom before the fry find them. Although these microorganisms are small and individually their decomposition does not cause much water pollution, they may have died by the thousands or even millions, so cumulatively their decomposition can pollute the aquarium water. One of the best ways to solve this problem is to keep several large ampullaria snails in the aquarium. These creatures are usually carrion feeders. A bottom layer of dead microorganisms is excellent food for them. They generally do not bother living fishes, even very tiny fry. These large snails do, of course, produce their own wastes, so the aquarium should be well filtered.

In an aquarium containing fry that are much smaller than the holes in the strainer on the end of the filter tube, filtration can be an acute problem, as the filter will suck in and trap the fry. This problem, however, is easily solved by using the right kind of filter for this situation. Sponge filters are ideal for this purpose. The pores in the sponge are smaller than the fry but large enough to admit bacteria and the free-floating proteinaceous matter that comes from decomposing plants and fish and snail wastes. Another easy solution is to use a standard inside box filter with a medium of fine sand or, preferably, filter floss, but do not install the lid on the box. This prevents the fry from being trapped in the filter. In fact, the filter actually becomes a good feeding ground for the fry, for microorganisms, brine shrimp nauplii, and other small food organisms will be drawn to the surface of the filter floss, where the fry can easily feed on them and then swim out of the filter.

As mentioned earlier, wild-caught masses of live food should be screened very carefully for predatory creatures that potentially are a threat to the fry. Hydra, often introduced with the catch, can be a severe problem. Also, cyclops nauplii that remain uneaten or are older when introduced into the tank can prey on the fry of labyrinth fishes.

When the young are about a week old, they can be switched to larger foods, such as brine shrimp nauplii, older cyclops nauplii, microworms, or finely ground dry foods.

There is another school of thought on the feeding of labyrinth fish fry. Some hobbyists do not have the room to raise large broods and are interested in raising only a few of the fry. The smaller labyrinth fish larvae cannot eat whole live brine shrimp nauplii, but they can and do tear them apart. By this method these fishes are actually capable of feeding on larger foods. Those individual fry that are naturally more aggressive than the others will get more food. Thus many of the fry will die of starvation, but a number of them will live and get a more than adequate amount of food. Using newly hatched brine shrimp nauplii or microworms as a first food for tiny fry eliminates not only the possibility of introducing predators into the aquarium but eliminates all the trouble of collecting, sorting, and cultivating these organisms.

The young of labyrinth fish species with sinking eggs—*Betta splendens* again being the exception—are much larger when they begin to swim free than the young of *Betta splendens* or any of the species with floating eggs. Also, the broods are much smaller, especially in the mouthbrooders. Because these young are larger, they can accept larger food organisms right from the start. For example, most of them can take newly hatched brine shrimp nauplii as their first food. Feeding the young of mouthbrooding labyrinth fishes is therefore not much of a problem, yet mouthbrooding species are still not bred regularly in the aquarium. This may be because the fishes are so rarely imported, in addition to the fact that they require special water conditions and are simply not very easy to breed. Some mouthbrooding species, such as *Betta taeniata* and the chocolate gourami, *Sphaerichthys osphre-*

menoides, have been bred in the aquarium, but not very often.

Culling

No matter which food is fed and no matter how well the brood as a whole eats (whether they are the fry of species with sinking eggs or floating eggs), there will always be a nonuniform growth rate among individuals of any one brood. This is because some individual fry are naturally more aggressive than others, as mentioned earlier, and they simply get more food. In raising a large brood, this problem is amplified as the fry grow, for those that have become larger by getting more food continue to outgrow the others, only even more so as the size disparity increases. Because of this problem and also because a few relatively small aquariums are generally more convenient than one huge one, the fry should be continuously sorted by size, with the larger fry kept in one aquarium and the smaller ones in another. This process is called culling, and it promotes a uniform growth rate among most of the fry.

The first culling should not take place for a few weeks, because the fry, not yet having fully developed labyrinth organs, are very sensitive to cool atmospheric air. Completion of labyrinth organ development can be discerned by watching to see when the fry begin to take gulps of air at the surface. Once this behavior occurs on a more or less regular basis, it is safe to move the fry for their first culling.

With Siamese fighting fish *(Betta splendens),* the culling technique just described applies to females all the time but to males for only about three months. About that time their natural instinct to fight with each other begins to develop, and the smaller fish not only starve but are also constantly picked on by the larger ones. The males can be kept together for as long as six or seven months, as long as their precariously balanced social hierarchy (pecking order) is not disturbed. In such a tenuous social order, the removal of one or two fish can trigger mass pandemonium in the aquarium as a new social hierarchy is established. Therefore, it is best to separate each male in its own container at the age of three to four months, before the social order really becomes established. This social hierarchy may only be an artifact of life in captivity. Nonetheless, it is a very real problem.

In most commercial hatcheries the males are kept separately, each in its own small container. This is also how these fish are stocked in retail stores. There is one alternative used in some commercial hatcheries in which numerous large broods are raised. In these establishments tranquilizers are added to the water so that many maturing males can be kept together in large pools or aquariums. This method is also used for shipping aggressive fishes in large batches. It is not a method recommended for hobbyists, for tranquilizers are difficult to come by, except in large commercial quantities, which are usually quite expensive. They are also not recommended because some degree of special knowledge is required for their proper use.

Water Changes

Continuous partial water changes aid the growth and overall well-being of the fry, because this keeps water pollutants well diluted, thus reducing the stress caused by the pollutants. In a large breeding aquarium, about one fourth of the water should be drawn off weekly and replaced with fresh water of the same temperature and chemical composition (pH and DH). While siphoning out the old water, decaying organic matter can be removed from the bottom of the aquarium, thus promoting a cleaner overall aquatic environment. In small aquariums containing large broods the water should be partially replaced at more frequent intervals, for example every two or three days. Some of the most robust fish come from aquariums in which the water is partially changed daily during the time they are young and growing rapidly.

SELECTION OF BREEDERS

Only fishes with the best body and fin form and coloration should be used as breeding stock. Most labyrinth fishes are sexually mature long before they have grown to full size. For those species that grow quite large it may, for reasons of space availability, be advantageous to breed them at less than full size. For most, however, better results will be achieved at full maturity for several reasons. By the time they are full grown most of the external features, such as body form and color, that are genetically predetermined to manifest themselves will have done so. From the array of variability the aquarist can then easily select the most desirable features.

A second reason for using mature breeders is that the various features of their highly ritualized behavior, after a few practice runs, becomes firmly fixed. Thus more eggs are generally laid, a greater percentage are fertilized, and a larger brood results. This can help the hobbyist in two ways, depending on the outcome of the second-generation offspring. There can be more variants of the trait being bred for from which to choose, or there can be greater numbers of individuals with exactly the desired trait.

In the selective breeding of the Siamese fighting fish, *Betta splendens,* many hobbyists prefer to breed younger fish rather than full-grown specimens. This is because the females are not usually injured as much during breeding and are less likely to be killed than they would be with full-grown males. If, however, large broods are desired, there are several ways to prevent the female from being badly injured or killed by a full-grown male. One is to simultaneously place several ripe females together with the male. These are not monogamous fish, and very often a male will breed with several different females before turning all his efforts toward brood tending. This distributes the total number of aggressive attacks over all the females, thus not subjecting any one female to enough abuse to cause serious injury. Another technique is to keep only one female in the tank at a time but keep changing females until a more compatible mate is found.

In selecting good breeders, disadvantageous traits must sometimes be accepted to get the desired traits of another feature. Disadvantageous traits, such as an odd or shorter fin shape, may suddenly appear when closely related fish are bred with one another (sibling-to-sibling or offspring-to-parent matings). Such a breeding scheme is known as inbreeding. Its deleterious effects are grossly exaggerated in much of the available aquarium literature. With careful selection, many breeding lines can be maintained for years without very many significant signs of degeneration. Furthermore, it often occurs that when two well-established breeding lines are crossed, the progeny are unusually robust and well colored in the first generation. This phenomenon, commonly known among professional plant and animal breeders, is called heterosis, or hybrid vigor. Sometimes, with a great deal of very careful selection, new improved breeding lines can be established from such a cross. This technique is often used by those breeders that have a bent for show trophies, but very often when these more robust offspring are mated together, the next generation reverts back to a normal appearance.

Armed with the knowledge that inbreeding is not always bad or is not always all bad, a hobbyist with no access to unrelated partners of a rare species can, with some degree of assurance, mate siblings to siblings without having to worry about ruining the fish. If one of the original partners of such a rare species is lost, breeding sibling to parent may not cause any significant degeneration for many years to come.

This information should not be interpreted to mean that haphazard inbreeding will not lead to problems, for inevitably, sooner or later, it will. Inbreeding should only be used as a temporary means of meeting a breeding goal, such as establishment of a particular trait. Unrelated breeding partners should be used whenever possible, for besides form and color, other traits, such as disease resistance, general vitality, fertility, fecundity, and general ease of breeding, are affected, usually adversely, by inbreeding.

DISEASES

A number of infectious fish diseases occur both in nature and in the aquarium. In nature, however, normal healthy fishes have more opportunity to remain healthy than they would if they lived in an aquarium. In even the most spacious aquarium, fish have much less space in which to live than they have in almost any wild habitat. Furthermore, most hobbyists' aquariums, more often than not, contain many more fish than should be kept in so small a space. The crowding itself stresses the fish. Adding to this stress are the effects of an excessive amount of pollution caused by an abundance of fish wastes and other kinds of decaying organic matter. This causes harsh changes in water chemistry, which irritate fishes' delicate skin, fins, and gills. These changes also upset the fishes' ionic balance. Moreover, these conditions deplete the oxygen supply by increasing the chemical oxygen demand, which is caused by the excessive number of oxidation reactions that take place in polluted water. The oxygen supply is also depleted by the increasing biological oxygen demand that results from the proliferation of microorganisms that consume organic matter and by the proliferation of disease organisms brought about by the weakened condition of the fish. Stress favors the development of all kinds of infectious diseases. The outbreak of infectious diseases following the long journey required for importation is most often the direct result of stress caused by the poor environmental conditions of the journey. The addition of new fish to an existing aquarium contributes to the spread of infectious diseases in two ways: the diseases are carried in with the new fish, and the addition itself temporarily stresses the old residents of the tank, by upsetting established social hierarchies and territories, thus making the old fish more susceptible to the newly introduced diseases. Fish are weakened by decreasing pH values and, contrarily, by increasing free ammonia when pH values are reversed. They are also weakened by an increase of nitrites and nitrates, compounds that naturally result from the normal metabolic processes of aerobic decomposer organisms in the tank. Diseases caused by improper feeding are common in labyrinth fishes, especially, as mentioned earlier, in the fry. Proper feeding can help prevent many malnutrition or deficiency diseases, but it cannot, in most cases, cure these diseases once they occur.

The best way to prevent diseases in labyrinth fishes is to practice good aquarium management. This includes frequent monitoring of water conditions (pH, DH, temperature, and so on); frequent partial water changes (including the simultaneous removal of uneaten food, dead fish, or other organic matter); sterilization of all equipment, such as nets and siphons, each time these items are used; and a quarantine period of two to three weeks for any new fish prior to their addition to the aquarium in question. Moreover, frequent checking of equipment, such as heaters, filters, and thermometers, for proper operation helps prevent disease outbreaks. Making sure all tankmates are compatible species can also help reduce disease outbreaks, as fewer injuries as a result of territorial and rivalry fights will occur.

The hobbyist who buys labyrinth fishes through the usual retail channels has no way of knowing how long the purchase was in quarantine before being offered for sale, or if it was quarantined at all. It therefore behooves hobbyists, no matter what they are told by the retailer, to quarantine new acquisitions for two to three weeks in a separate aquarium, where they can be closely watched for any signs of disease. This quarantining procedure vastly reduces the chances of introducing diseases into the aquarium. If a disease is detected in the quarantine tank, it can be properly treated there without exposing the other healthy fish to the stresses caused by harsh medications, or the fish can be returned to the retailer from whom it was purchased. This method prevents rapidly spreading diseases, such as ich, velvet, fin rot, a disease caused by numerous bacterial and fungal species, and mouth fungus, a highly contagious fungal disease that attacks the mouth, throat, and gill area of fish. Quarantine, however, is not an effective method for prevention of the diseases that spread more insidiously, such as fish tuberculosis or hexamitiasis (hole-in-the-head disease), unless, of course, the quarantined fish is in an advanced stage of such diseases. Fishes bearing such advanced diseases, however, should be identified on the premises of the retailer and not purchased at all.

There are numerous effective medications available for treatment of infectious and parasitic diseases. Treatment is usually by means of short-term or long-term baths. Treatment with medicated food would, in theory, work better than baths for many diseases, but in fact it usually does not work, because by the time most internal diseases are detected they have progressed far enough to cause the fish to lose its appetite; thus the medicated food cannot be consumed.

Most disease treatments should not be carried out in the community aquarium or in the breeding aquarium, as most of them also eradicate the beneficial microorganisms that live in the tank and help maintain its chemical and biological balance. Although many of these medications are available from pharmacies, most are by prescription only and are very expensive. Most of the chemicals used to treat diseases are also available from chemical or biological supply houses, but only in large quantities—much larger than any aquarist will ever need—and such large

quantities are also very expensive. Most effective medications are available from tropical fish stores in reasonable quantities and at reasonable prices. They are also packaged in ready-to-use concentrations. Each manufacturer provides on the label directions for dosages and lists of diseases for which the medication is effective. With few exceptions, the manufacturer's directions should be followed exactly.

DISEASES CAUSED BY PROTOZOA

Counted among the protozoa are a number of unicellular organisms, such as flagellates, ciliates, amoebas, and sporozoa. Only a few of them cause diseases in fish. These diseases are discussed in the sections that follow.

Velvet Disease

Oodinium pillularis, the agent that causes velvet disease, was first found in *Colisa* species; therefore, this disease is sometimes referred to as the colisa disease. *Oodinium pillularis* is not restricted to labyrinth fishes, although labyrinth fishes and killifishes are more susceptible to the disease than are most other kinds of exotic fishes. Infected fish sometimes rub themselves against various objects, which apparently relieves the irritation caused by these parasites. In this disease, large areas of the skin become covered with a dustlike velvety-appearing coating of cysts, which individually are not visible to the unaided human eye. In heavily infected fish this coating of cysts covers the entire body and the gills. These parasites can also live on the mucosa of the stomach and intestines. In a microscope the parasites attached to the skin appear to be pear-shaped. As soon as an individual parasite is completely developed, it separates from the fish and sinks to the bottom of the aquarium. After passing through an intermediate developmental stage, the infective cell begins to divide as it lies on the bottom, and thirty-two to sixty-four flagellated swarmers emerge and swim free, each seeking a new host. The flagella are shed as soon as the parasite attaches to a host. If a host is not found within twenty-four hours, the parasite dies. After a long-term siege of velvet, tiny craterlike depressions appear on the skin where the parasites were attached. This parasite can also penetrate deeply into gill tissue, so erratic breathing may also be one of the symptoms.

The best way to eradicate these parasites is to treat the water with a chemical that will kill them in the free-swimming swarmer stage. Any medication strong enough to kill this parasite in its encysted stage is also likely to kill the host fish. The drug of choice is a pharmaceutical dye called acriflavine (trypoflavine), which is available in both liquid and powdered form from most tropical fish stores. The treatment should be continued for no fewer than ten days, to be certain that all the parasites have passed through the swarmer stage.

Costiasis

Costiasis is a skin-destroying disease caused by *Costia* species. These flagellates weaken labyrinth fishes, especially young fish that are kept in poor aquarium conditions. Adult fish are most subject to the infection after having been transported or when the aquarium water suddenly becomes too cool. Not only does the temperature drop weaken the fish, but it causes a proliferation of the parasite, which happens to thrive in cooler water.

Costiasis can be recognized by the appearance of a fine darkish patchy coating, especially on the soft tissue of the fins. Heavily attacked fin and skin areas are often suffused with exposed blood vessels, and the fish often shake from side to side. *Costia neatrix* parasitizes epidermal cells and destroys them. Reproduction in the flagellate occurs by cell division. The pear-shaped body is equipped with two flagella with which the parasite propels itself. Without a host it is viable for only about an hour. Since this parasite dies at temperatures over 86°F (30°C), and since

labyrinth fishes thrive quite well at this temperature, the organisms can be killed by a corresponding temperature increase for a period of about three to five days. If eradication does not come about after this treatment, it is possible that the parasite is not *C. neatrix* after all. Ciliated parasites, such as *Chilodonella* species and *Trichodina* species, produce the same symptoms as *C. neatrix*. All these skin-destroying parasites can be eradicated by a long bath in acriflavine in a dosage of 1 g to 25 gallons (100 liters) of aquarium water for at least four days.

Ich

Ichthyophthirius, usually known by its shortened form ich, is caused by a ciliated protozoan called *Ichthyophthirius multifiliis,* and it frequently attacks labyrinth fishes. This parasite, which attacks the skin and gills of the fish, typically appears as a covering of small white spots, each about the size of a table salt crystal; hence the disease is sometimes known as white spot disease. The white spots are encysted feeding parasites. After a parasite has dropped off the body of the fish, it reproduces by simple division, releasing hundreds of new swarmers that seek out new hosts. Since these swarmers are so numerous and can live without a host for up to forty-eight hours, infection is quite likely to occur, if conditions are right. Weakened, chilled fish *are* the right conditions. The parasites in the gills escape even the most thorough check and probably are the cause of suddenly occurring or recurring epidemics of this disease, especially in the stocks of importers, wholesalers, and retailers. As in velvet disease, any chemical strong enough to kill the parasite in its encysted stage is also strong enough to kill the host fish as well. The drug of choice is malachite green chloride, which kills only the swarmers, not the hosts. The procedure for treatment is about the same as it is for the treatment of velvet disease, a ten-day bath, but in malachite green. Elevation of the water temperature to about 86°F (30°C) also helps eradicate this disease and can be used in conjunction with the malachite green treatment. The ten-day treatment alone often eradicates this pest without the use of chemicals; thus treatment can be administered right in the community aquarium.

DISEASES CAUSED BY FUNGI

Several species of fungi can produce infections in labyrinth fishes, both external and internal infections.

External Fungal Infections

Infections by the fungi *Saprolegnia* and *Achyla* are not uncommon in labyrinth fishes. They are usually secondary infections that follow infections caused by bacteria or other organisms that damage tissue. Spores of these fungi are present in almost every aquarium but normally pose no danger to the fish. Only severely weakened or externally damaged fishes become infected. These fungi can be recognized as white cottony tufts at the infection site or on the fins as a whitish ragged edging. The most common remedy is exposure to a fairly strong saline solution, but not all labyrinth fishes can tolerate this treatment. For those that can, the treatment should begin with a fifteen-minute bath at a strength of 2 tablespoons of table salt per gallon of water. The fish should be watched carefully for signs of distress, such as erratic breathing or erratic swimming motions. The treatment is more effective if the salt concentration is increased. If the fish react adversely to the first treatment, however, the solution should be gradually diluted until normal behavior resumes. But if they can tolerate more salt, gradually build up the concentration to a maximum of 2 tablespoons more per gallon of water, watching carefully for signs of distress. Once the strongest tolerable concentration is found, the treatment should be repeated as often as possible but in no case fewer than three times a day. This daily treatment should be continued until the fungus is gone.

Another effective treatment is to swab the infected area with a disinfectant, such as Mercurochrome®, using a cotton swab. For this treatment the fish can be gripped in a wet net. The fish should be held out of the water for thirty to sixty seconds after the treatment to give the disinfectant time to work before it is washed off in the aquarium water. The small amount of disinfectant that will wash into the water will do no harm to the fish or the plants. During the treatment period and for several days afterward, the fish should be kept in an isolation tank to help prevent the fungal infection from returning and to allow the fish a few days of rest to regain its strength before once again having to contend with its tankmates. The infection will be conquered sooner if both the Mercurochrome® and the salt bath treatments are used simultaneously, with the Mercurochrome® swabbing immediately following each salt bath. For those species that cannot tolerate the salt bath at all, the Mercurochrome® treatment alone, repeated several times a day, will help.

Some labyrinth fishes are occasionally attacked by columnaris disease, the symptoms of which closely resemble those of external fungal infections, especially of the mouth. This disease does not respond to the treatments used for fungal infections. The description and treatment of this disease is covered later on in this chapter.

Internal Fungal Infections

Several fungus species are known to infect the internal organs or other internal tissues of fish. The best known of these is *Ichthyophonus hoferi*, which can invade all organs. Specific symptoms vary somewhat according to the organ or organs affected, but generally include loss of color, loss of vigor, loss of appetite, and severe wasting. Sometimes areas of the skin may have a roughened appearance. There is no known effective treatment. The victims gradually waste away and die; sudden deaths and deaths of epidemic proportions do not occur.

The symptoms of *I. hoferi* cannot be distinguished from those of piscine tuberculosis, a bacterial disease. Only a thorough microscopic examination of necropsied animals can distinguish the two diseases. Since fish tuberculosis is also fatal, the aquarist is advised to destroy infected fishes as quickly and painlessly as possible and carefully dispose of their carcasses by incinerating them or placing them in the trash, wrapped tightly in a plastic bag. *Never* flush them down the toilet, because the disease organisms can still be alive and infect fishes living in local waters.

Following a bout with this disease, the aquarium should be dismantled and thoroughly disinfected along with all the items it contains except the gravel and live plants. The plants should be carefully discarded and the gravel may be disinfected by soaking it in boiling water (not in the aquarium) for about fifteen minutes. The tank, heater, filter, plastic plants, and other equipment can be disinfected by soaking in a strong solution of formalin or methylene blue for about twenty-four hours. Alum in powdered form is available from most pharmacies in small quantities. It is an effective disinfectant that does not give off noxious fumes, nor does it discolor the aquarium or equipment. It is most important that all this equipment be thoroughly rinsed in cool or tepid (not hot) water before being put back into use. Porous rocks and driftwood should be boiled along with the gravel, rather than soaked in any of these disinfectant solutions.

DISEASES CAUSED BY BACTERIA AND RELATED ORGANISMS

A large number of bacterial diseases occur in fishes in general but these diseases can only be specifically identified in specialized diagnostic laboratories or by hobbyists with a good microscope and a thorough knowledge of diagnostic techniques.

Fish Tuberculosis

Fish tuberculosis generally does not appear spontaneously in a well-maintained aquarium. Labyrinth fishes imported from Southeast Asian wholesale breeders sometimes carry this disease; fishes purchased from commercial American breeders rarely do. Whether the disease is actually fish tuberculosis or *Ichthyophonus hoferi* is merely an academic question. Either way, the fish must be destroyed and disposed of.

sical (have bloated bellies). Holes in body tissue, especially along the flanks, may develop, and the eyes may protrude markedly. No effective treatment is known. Diseased fishes should be humanely destroyed and carefully disposed of.

Fish tuberculosis generally does not appear spontaneously in a well-maintained aquarium. Labyrinth fishes imported from Southeast Asian wholesale breeders sometimes carry this disease; fishes purchased from commercial American breeders rarely do. Whether the disease is actually fish tuberculosis or *Ichthyophonus hoferi* is merely an academic question. Either way, the fish must be destroyed and carefully disposed of.

Columnaris Disease

In labyrinth fishes this disease is prevalent in the Siamese fighting fish, *Betta splendens,* but it is seen in other species as well, especially livebearers (Poeciliidae). It is caused by the myxobacterium *Chondrococcus columnaris*. It is seen as large whitish spots, especially around the mouth, where it is often mistaken for mouth fungus. On infected skin areas, the spots spread out and eventually cause the skin in the infected areas to become ulcerated. Scale loss may be observed. If not quickly controlled, the myxobacterium eventually invade the subdermal tissue, permanently destroying muscle fibers. The antibiotic Furanace™ is the drug of choice for treating this disease. This drug is commercially available from most tropical fish shops, packaged in dosages suitable for use by the aquarist.

It should be administered according to the manufacturer's directions, and treatment should continue for at least three days. Treatment should not be administered in the community aquarium or the breeding aquarium, as Furanace™ is a wide-spectrum antibiotic that targets many bacterial species, including the beneficial bacteria that help maintain the delicate biological and chemical balance of the aquarium. If treatment must be administered in the community or breeding aquarium, half of the water should be replaced following treatment, and the remaining water should be filtered through a bed of high-grade activated charcoal.

Other bacterial infections, such as bacterial skin and bacterial gill disease, respond equally as well to treatment with Furanace™, for this drug is easily absorbed into fishes' bodies through fine skin capillaries and through the fine capillary network of the gills.

PARASITIC INFESTATIONS

Labyrinth fishes are generally quite resistant to parasitic crustaceans, such as fish lice (*Argulus* sp.); nematode worms, such as *Camallanus* species; or copepods, such as *Ergasilus* species. Gouramis, however, especially large ones, are occasionally attacked by species of the crustacean *Lernaea*. This parasite bores deeply into the fish's skin and betrays its presence by the large egg sacs of the females, which hang wormlike from the body of the fish. *Lernaea* and most other large parasites can be eradicated with trichlorfon, which is the generic name for a drug called Dylox®. The drug is also sold under several other trade names. This drug, originally marketed as a cattle insecticide, is now packaged for aquarium use as the primary ingredient or one of a few secondary ingredients in several different aquarium medication preparations. Instructions for their use are printed on the packages.

DISEASES OF INTERNAL ORGANS

Fish diseases of internal organs cannot normally be treated. One such disease, unfortunately rather prevalent in labyrinth fishes, is referred to by the comprehensive term "adiposis" (steatosis), or fatty degeneration of the viscera. The fatty degeneration of internal organs, especially of the ovaries and liver, can interfere with reproduction and can lead to complete sterility. In liver tissue fatty deposits are fatal, for they interrupt liver functions, which are crucial for survival. Specifically responsible for these fatty deposits is the accumulation of unsaturated fatty acids. Adiposis may be one of the most frequent "unknown" causes of death in aquarium-kept labyrinth fishes. The disease is brought about by improper nutrition and can be prevented only by proper nutrition.

POISONINGS

In labyrinth fishes poisoning can be caused not only by poisonous substances in the water or in the food but also by poisons from the air, which act on the labyrinth organs of these fishes. Among these poisons are tobacco smoke, insecticides, and vapors from solvents. Poisonings of labyrinth fishes by solvents evaporated from paints being applied in the home are known. Before extensive painting is done, the aquarist should remove the aquariums or the fish from the rooms in which such work is to be done. Also dangerous to labyrinth fishes is excessive exposure to pure oxygen. Many a labyrinth fish has succumbed to the influence of pure oxygen on the labyrinth organ, particularly during transportation by the breeder or wholesaler, when the fish are shipped in plastic bags filled with pure oxygen. As long as the fish being shipped are species that only breathe oxygen dissolved in water, via their gills, the use of oxygen is absolutely safe. However, in all fishes that possess accessory breathing organs, serious damage to the mucosa of the labyrinth organ can result from contact with pure oxygen. Based on my experience, *Anabas* and *Trichogaster* species are especially sensitive to pure oxygen. Rarely do the damages suffered by labyrinth fishes exposed to pure oxygen lead to immediate death. Usually the affected fish die three to ten days after their journey. Necrosis of the labyrinth organ can only be detected during necropsy, once the fish has died. The direct cause of death is asphyxiation.

TOP LEFT: A pair of paradise fish, *Macropodus opercularis,* under their bubblenest. The male is the fish on the left.

TOP RIGHT: A giant gourami, *Colisa fasciata.* This species is widely distributed in India. It is variable in color.

CENTER LEFT: A male honey gourami, *Colisa chuna,* in its breeding colors.

CENTER RIGHT: A male of the true-breeding red color form of the dwarf gourami, *Colisa lalia.*

BOTTOM LEFT: A male of the normal or wild-type color form of the dwarf gourami, *Colisa lalia.*

BOTTOM RIGHT: A pair of thick-lipped gouramis, *Colisa labiosa.* The male is the fish on the left.

DESCRIPTIONS OF SPECIES

In the following species descriptions, the labyrinth fishes are divided into Asian and African species, with the Asian species discussed first. The family Anabantidae, to which both Asian and African species belong, forms the transition between the two groups. The genus-and-species-rich family Belontiidae is divided into five subfamilies. Of these, two, Ctenopinae and Sphaerichthyinae, produce sinking eggs; the other three, Trichogasterinae, Belontiinae, and Macropodinae, produce floating eggs.

Within the families and subfamilies listed, the genera and the species within each genus are arranged alphabetically. In this classification, the species are separated into those with floating eggs and those with sinking eggs. The species of the genus *Betta* are arranged into species that build bubblenests and those that are mouthbrooders.

Characteristics common to all the species of each genus are summarized under each genus name. The species descriptions contain the common and scientific names and also the most important synonyms (names that are not valid scientifically but that have been in common use). Knowing these synonyms, including the appropriate author's name, often facilitates finding information on a species in older literature. The species descriptions also contain statements about the maximum size and the appearance of each

species as well as information about the geographic distribution and ecology. Suggestions about the best conditions for keeping them in the aquarium, pointers on feeding, and details about breeding are also given for each species. Popular mutant forms and true-breeding color varieties are described immediately following the description of the species.

The descriptions of the labyrinth fishes are followed by descriptions of the pikeheads (Luciocephalidae) and of the snakeheads (Channidae). In the snakehead descriptions, emphasis is on the smaller species, which not only can be kept in the aquarium but can also be bred.

Described Genera and Species

Asian Labyrinth Fishes

Family Belontiidae
Subfamily Belontinae
Belontia
    B. hasselti (Cuvier, 1831)
    B. signata (Günther, 1861)

Subfamily Macropodinae
Macropodus
    M. chinensis (Bloch, 1790)
    M. concolor Ahl, 1937
    M. opercularis (Linnaeus, 1758)

Subfamily Trichogasterinae
Colisa
    C. chuna (Hamilton & Buchanan, 1822)
    C. fasciata (Bloch u. Schneider, 1801)
    C. labiosa (Day, 1878)
    C. lalia (Hamilton & Buchanan, 1822)
Trichogaster
    T. leeri (Bleeker, 1852)
    T. microlepis (Günther, 1861)
    T. pectoralis (Regan, 1910)
    T. trichopterus trichopterus (Pallas, 1777)
    T. trichopterus sumatranus Ladiges, 1933

TOP: A pair of pearl gouramis, *Trichogaster leeri,* under their bubblenest. The female shows her readiness for spawning by pushing her snout into the side of the male.
BOTTOM: In spite of its large size, the relatively quiet demeanor of the moonbeam gourami, *Trichogaster microlepis,* from Thailand, makes this fish suitable for the company of somewhat smaller fishes.

Descriptions of Species

Subfamily Sphaerichthyinae
Parasphaerichthys
   P. ocellatus Prashad und Muckerji, 1929
Sphaerichthys
   S. acrostoma Vierke, 1979
   S. osphromenoides Canestrini, 1860
   S. osphromenoides selatanensis Vierke, 1977
   S. vaillanti Pellegrin, 1930

Subfamily Ctenopinae
Betta (Schaumnestbauer)
   B. bellica Sauvage, 1884
   B. coccina Vierke, 1979
   B. fasciata Regan, 1909
   B. foerschi, Vierke, 1979
   B. imbellis Ladiges, 1975
   B. smaragdina Ladiges, 1972
   B. splendens Regan, 1909
Betta (Maulbrüter)
   B. anabatoides Bleeker, 1850
   B. balunga Herre, 1940
   B. brederi Myers, 1935
   B. macrostoma Regan, 1909
   B. picta (Cuvier u. Valenciennes, 1846)
   B. pugnax (Cantor, 1850)
   B. rubra Perugia, 1893
   B. taeniata Regan, 1909
   B. unimaculata (Popta, 1905)
Ctenops
   C. nobilis McClelland, 1844
Malpulutta
   M. kretseri Deraniyagala, 1937
Parosphromenus
   P. deissneri Bleeker, 1859
   P. filamentosus Vierke, 1981
   P. paludicola Tweedie, 1952
   P. parvulus Vierke, 1979
Pseudosphromenus
   P. cupanus (Cuvier u. Valenciennes, 1831)
   P. dayi (Köhler, 1909)
Trichopsis
   T. pumilus Arnold, 1936
   T. schalleri Ladiges, 1962
   T. vittatus (Cuvier u. Valenciennes, 1831)

Family Helostomidae
Helostoma
   H. temminckii (Cuvier u. Valenciennes, 1831)

Family Osphromenidae
Osphronemus
   O. goramy Lacépède, 1802

Family Anabantidae (Asian species)
Anabas
   A. testudineus (Bloch, 1795)

Family Anabantidae (African species)
Ctenopoma
   Brood-caring species
   C. ansorgii (Boulenger, 1912)
   C. congicum Boulenger, 1887
   C. damasi (Poll, 1939)
   C. fasciolatum (Boulenger, 1899)
   C. nanum Günther, 1896
   Non-brood-caring species
   C. acutirostre Pellegrin, 1899
   C. argentoventer (Schreitmüller, Ahl, 1922)
   C. kingsleyae Günther, 1896
   C. maculatum Thominot, 1886
   C. multispinis Peters, 1844
   C. muriei (Boulenger, 1906)
   C. nigropannosum Reichenow, 1875
   C. ocellatum Pellegrin, 1899
   C. oxyrhynchum (Boulenger, 1902)
   C. pellegrini (Boulenger, 1902)
   C. petherici Günther, 1864
Sandelia
   S. bainsii Castelnau, 1861
   S. capensis (Cuvier u. Valenciennes, 1831)

Pikeheads (Luciocephalidae)
Luciocephalus
   L. pulcher (Gray, 1830/34)

Snakeheads (Channidae)

Asian species:
Channa
   C. lucia (Cuvier u. Valenciennes, 1831)
   C. marulia (Hamilton & Buchanan, 1822)
   C. melasoma (Bleeker, 1851)
   C. micropeltes (Cuvier u. Valenciennes, 1831)
   C. orientalis Bloch und Scheider, 1801
   C. pleurophthalma (Bleeker, 1850)
   C. punctata (Bloch, 1793)
   C. striata (Bloch, 1797)

African species:
   C. africana (Steindachner, 1879)
   C. obscura (Günther, 1861)

Asian Labyrinth Fishes

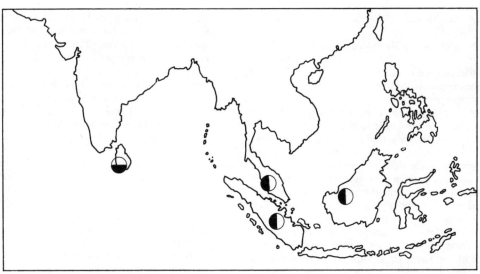

Distribution map of the genus *Belontia*

◐ *Belontia signata*

◑ *Belontia hasselti*

Asian Labyrinth Fishes

Family Belontiidae,
Subfamily Belontiinae,

Genus *Belontia*

The genus *Belontia* was established by Myers in 1923. It contains two species that bear the somewhat misleading common name "island macropodes" since they are neither closely related to the species of the genus *Macropodus* nor do they occur exclusively on islands. In addition to occurrences on Sri Lanka (*B. signata*) and Borneo and Sumatra (*B. hasselti*), the latter species also occurs on the Malay Peninsula (i.e., the Asiatic mainland).

Belontia species are relatively large fish whose appearance resembles cichlids more than it does labyrinth fishes. Keeping them in community aquariums together with smaller fishes is not advised. They are especially aggressive during the spawning period, particularly toward their own kind. Socialization is recommended only with fishes of corresponding size.

In nature, species of *Belontia* occur mainly in running waters but they have been found in small bodies of standing water. The labyrinth organ is relatively small and not very convoluted. The fish are dependent on their labyrinth organ for oxygen only to a limited extent. They can remain submerged for a long time without having to come up for air. At night, they often rest on the bottom, on their sides, and assume a different color pattern.

As a rule, a bubblenest is not built. If one is, it is usually only after spawning is complete. After the territory has been established, mating takes place near the water surface. During the embrace, the male arches his body over the female and turns her belly up. The released eggs are fertilized as they float to the surface. The eggs are relatively large and have a diameter of 1.2 to 1.5 mm. Between each individual mating, the eggs are gathered together mainly by the male, although sometimes the female participates in this activity. When apparent danger presents itself, the males take the eggs, a few at a time, into their mouths and carry them to a different location. In full-sized adult fish, the number of eggs in one spawning varies from 400 to about 600.

Belontia hasselti (Cuvier and Valenciennes, 1831)
COMMON NAME: Honeycombed comb-tail, archipelago macropodus
SYNONYMS: *Polyacanthus hasselti* Cuvier and Valenciennes, 1831; *Polyacanthus kuhli* Bleeker, 1845. Two variant forms were described by Bleeker as *Polyacanthus einthovenii* Bleeker, 1851 and *Polyacanthus helfrichi* Bleeker, 1855.
LENGTH: Up to 7½ in. (19.5 cm).

FIN RAY COUNTS: (D, XVI–XX/9–13; A, XV–XVII/11–13.

DISTRIBUTION: Malay Peninsula, Borneo, and Sumatra. It is said to occur also on Java. Found mainly in slowly running, clear waters. The first importations of live specimens occurred in 1966, simultaneously by several importers.

DESCRIPTION: Males are more slender than females. The unpaired fins are larger in the male; the posterior ends of the fins are longer and more pointed. *Belontia hasselti* has a typical spot pattern in which the individual spots have the shape of honeycombs. These spots form a uniform pattern that extends over the tailfin, the caudal peduncle, and the soft parts of the dorsal and anal fin. The primary color is gray-brown, occasionally with a slightly green-brown hue. The scales have darker edges. The markings change during periods of darkness. In the dark, a striped pattern appears that consists of dark, irregular horizontal stripes. The rest of the body becomes yellowish. These fish assume a similar color pattern during the mating season. During this time, the body color of the female is usually a little lighter than normal.

MAINTENANCE AND BREEDING: No special care is needed. These fish can be bred in almost any type of water. They are suitable only for large aquariums of 50 g (200 liters) or more, because they grow to a very large size as adults. High temperatures are best; from 77°F (25°C) up to about 83°F (28°C). For breeding the aquarium should contain numerous hiding places.

Belontia hasselti is primarily a carnivore but also eats some plants. Spawning normally occurs without the presence of a bubblenest, and the eggs are guarded by both parents. The eggs hatch after twenty-four to thirty hours of incubation at a temperature of about 83°F (28°C). The fry begin to swim freely three days later and can immediately be fed brine shrimp (nauplii) both live and frozen or properly screened live food from outdoors.

Belontia signata (Günther, 1861)
COMMON NAME: Combtail, Ceylon macropodus.
SYNONYMS: *Polyacanthus signatus* Günther, 1861.
LENGTH: Up to 5½ in. (14 cm).
FIN RAY COUNTS: D, XVI–XVII/9–10; A, XIII–XIV/10–12.
DISTRIBUTION: Several different geographic races occur in Sri Lanka, mainly in rivers but also in smaller bodies of water. This species was first imported to the West in 1933. Since then it has been imported irregularly.
DESCRIPTION: Two basic types can be distinguished in this species. One has a plump form with a high back, with its body length not exceeding three times its body height. Its primary coloration is reddish. This form might have been the basis of the original description by Günther. The second is a more slender type in which the predominant color is blue; it has a distinct blue spot on the base of the pectoral fin. The rays of the caudal fin extend beyond the soft tissue, and the tips of the dorsal fin and anal fin are longer than those of the more stocky type. There are probably transitional forms between the two types. Based on my own experience, the fish of the first type are much more aggressive than the more slender forms, which were imported later. Some fish show gray-brown or blue-red spots in their spawning coloration; these probably have nothing to do with sexual differences.

The males have a more slender body and more pointed dorsal and anal fins. Fully mature females are conspicuous by their greater body girth. It can be difficult to distinguish the sexes in immature fish.

MAINTENANCE AND BREEDING: *Belontia signata* requires no special water conditions, but it does require fairly high temperatures, about 77 to 86°F (25 to 30°C). It can be fed now and then with small earthworms in addition to other live food. Boiled fish is also readily accepted.

Small aquariums should not be used for breeding. The males establish fairly large territorial areas and sometimes do not even tolerate females in the territory for spawning purposes. Usually no structured bubblenest is built, but sometimes fragments of a nest are seen. After spawning is complete the female is usually not tolerated in the vicinity of the eggs, yet it has been reported in some pairs that the female does participate in brood care. The eggs hatch after about thirty hours at a temperature of 79 to 82°F (26 to 28°C). The young fish swim freely on the fourth or fifth day after spawning and are then about 5 to 6 mm long and can be fed right away with newly hatched brine shrimp nauplii.

Subfamily Macropodinae,

Genus *Macropodus*
The genus *Macropodus,* established by Lacépède in 1802, consists of three species as recognized by modern taxonomists. The species *M. cupanus* and the subspecies *M. cupanus dayi,* which were temporarily classified in this genus, have again been classified in the genus *Pseudosphromenus,* which had already been established for these fish in 1879 by Bleeker. In this earlier work the former subspecies *M. c. dayi* was recognized as a full species.

Representatives of the genus *Macropodus* are widely distributed in all of East and Southeast Asia, occurring from Korea in the north to Vietnam in the south. In

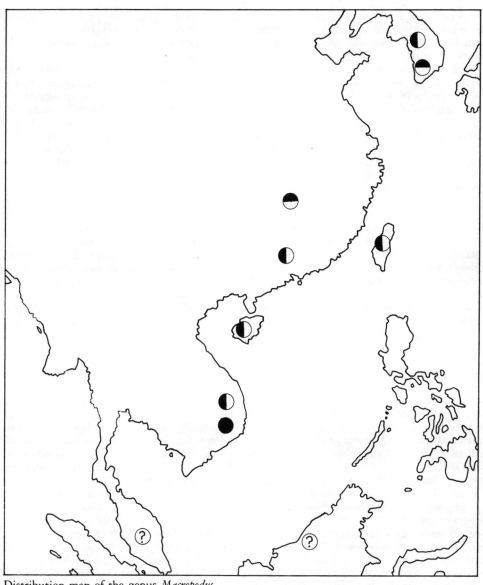

Distribution map of the genus *Macropodus*

◑ *Macropodus opercularis*

◓ *Macropodus chinensis*

⑦ alleged occurrence of *Macropodus concolor*

probable occurrence of ● *Macropodus concolor*

addition, occurrences are known on Taiwan, Hainan, and also on the Japanese Riukiu Islands, where it may have been introduced rather than being indigenous. The species *M. concolor* may also exist in Borneo. Occurrences of this species in North Borneo (Sarawak) can be traced back to releases by humans.

The three species of *Macropodus* are closely related and can be crossbred with each other. Hybrids resulting from such crossbreeding, however, often show a clearly lowered fertility rate and a very poor percentage of viable young. These hybrids are rarely completely sterile. According to H. Schwier, who published a paper about species hybrids within the genus *Macropodus*, the hybrid males are said to be sterile but the females are fertile. At present this cannot be tested with *M. chinensis* as one of the parental partners, because no *M. chinensis* are available. The information from Schwier apparently does not apply to the offspring of crosses between *M. opercularis* and *M. concolor*. It is known that hybrids from crosses between *M. opercularis* males and *M. concolor* females, and vice versa, in both backcrossings with partners of the original species and with each other, are fertile. According to Schaller (personal communication), hybrids from such crosses can be mated exclusively with each other for generations. In most of these hybrids; however, fertility problems do arise, but some individuals several generations removed from the original hybridization have proved to be completely fertile. In their external appearance the hybrids seem to fall somewhere between *M. opercularis* and *M. concolor*. Their intermediate appearance is maintained through many generations, and a reversion back to the original forms has not occurred. Most hybrids have the same gill cover spot as *M. opercularis*, but they have red pelvic fins, as in *M. concolor*. Older hybrid males often have extended caudal fin rays that are similar to those of *M. concolor*.

An interesting observation in crossbreeding attempts between *M. opercularis* and *M. concolor* is that better results are achieved in combining a *M. opercularis* male with a *M. concolor* female than vice versa. The cause may lie in the fact that the bright blue spot on the gill cover of the *M. opercularis* male functions as a sign stimulus for the *M. opercularis* female, and since this spot is missing in the *M. concolor* male, normal courtship is interrupted. It can often take weeks until mating actually occurs. If mating does finally take place, spawning may terminate before the female releases all the ripe eggs. In the reversed mating combination (i.e., *M. opercularis* male and *M. concolor* female), the mating prelude and the matings take place successfully.

In spite of their seemingly close relationship, one is probably justified in considering all three *Macropodus*

species as valid. There are differences in the values of the fin ray formula and considerable differences in the coloration and markings between *M. opercularis* and *M. concolor*, even though they are quite similar in body shape. Possibly further adding to the distinction between these species are ethological differences, such as different sign stimuli in mating behavior. On the other hand, the seemingly close relationship of the *Macropodus* species explains many of the contradictions in older ichthyological literature. For example, the renowned ichthyologist G. A. Boulenger (1858–1937) thought *M. opercularis* was actually a domesticated form of the wild species, which was possibly *M. chinensis*. Professor E. Ahl described *M. concolor* originally as a subspecies of *M. opercularis*. Other scientists even claimed to see a melanistic form of *M. opercularis* in this species.

This example shows how problematical is the borderline between species and subspecies. We will encounter this question again in other genera of labyrinth fishes. There is also another problem in defining these species: many of the large areas in which these species occur have for political reasons been inaccessible to research for decades. Ichthyological field studies have been practically impossible in such countries as Vietnam, Laos, Kampuchea, Burma, China, and certain parts of Indonesia.

In nature, *Macropodus* species occur in all kinds of waters. In Southern China and Vietnam they are, in some localities, the predominant fishes in flooded rice paddies. They possess a well-formed labyrinth organ and can survive for several weeks in mud-filled and partially dried-up irrigation ditches. The northernmost forms are extremely tolerant of the colder climate and can temporarily survive exposure to temperatures of about 40°F (5°C). Their food probably consists mostly of insect larvae and insects landing on the water surface. The males are industrious bubblenest builders that anchor their nests to a firm surface spot such as an emergent plant stalk or leaf. If surface plants or emergent surface leaves are available, they often build their nests under them. Most males are not very aggressive toward the females during establishment of their territories and bubblenest building. Most of the time the female is able to remain fairly close to the nest-building male without being chased away. There have been many reports of female *M. opercularis* participating in bubblenest building. I have not observed this firsthand, but I have seen females help defend their territory when the pair was kept in an aquarium together with other fishes.

Macropodus species are hardy, undemanding fish. Some species may generally be ill-tempered toward other fishes, but there are big individual differences in this behavior. Such differences in temperament can also be seen between strains. For example, the popular albino *M. opercularis* is generally much more mild-mannered than is the normal or wild form toward other fishes.

Reproduction occurs in *M. opercularis* at a temperature as low as 68°F (20°C), but at that temperature there is a considerable delay in embryonic and larval development. It is therefore recommended that all *Macropodus* species be bred at water temperatures of 77 to 83°F (25 to 30°C). At these temperatures the eggs hatch in about thirty-five hours, and after another three to four days the larval fish begin to swim free. Initially they stay close to the water surface and, if need be, can even be fed finely ground dry food. Better results will be achieved, however, by using small microorganisms, such as *Paramecium,* as their first food. After five to six days of feeding on these organisms they will be large enough to take freshly hatched brine shrimp nauplii. All *Macropodus* species willingly spawn even in very small aquariums. Raising a large number of young fish to a salable size, however, requires an aquarium of at least 20 to 30 g (80 to 120 liters).

Macropodus chinensis (Bloch, 1790)

COMMON NAME: Round-tailed macropodus
SYNONYMS: *Chaetodon chinensis* Bloch 1790; *Polyacanthus chinensis* Cuvier and Valenciennes 1831; *Macropodus ctenopsoides* Brind, 1915; *Macropodus chinensis* Myers 1925. In addition, Günther and Regan described this species as *Polyacanthus* or *Macropodus opercularis,* respectively, apparently based on the belief that this fish represents the wild form of the paradise fish.
LENGTH: 3 in. (7 cm), females somewhat smaller.
FIN RAY COUNTS: D, XIV–XVIII/5–8; A, XVIII–XX/9–12.
DISTRIBUTION: Widely distributed from Korea to Southern China. First imported to the West in 1914. Never imported or kept regularly in the aquarium trade.
DESCRIPTION: Fin tips of the dorsal and anal fins in the male are long and pointed, with an iridescent greenish color on the soft tissue between the fin rays. In contrast to the other species of *Macropodus,* the round-tailed *Macropodus* has a rounded tailfin. The tips of the dorsal and anal fin in the male protrude noticeably beyond the caudal fin. The background color of both sexes is yellow-brown to green-brown, and a number of irregular dark brown diagonal stripes appear on the body. The male is more intensely colored than the female.
MAINTENANCE AND BREEDING: Since this species has not been kept as an aquarium fish for several decades, recent breeding results are not available. According to the available information, reproduction in this species probably does not differ from that of other *Macropodus*

species. It is hoped that a renewed importation of this species is not long in coming.

Macropodus concolor Ahl, 1937

COMMON NAME: Black *Macropodus*
SYNONYMS: *Macropodus opercularis concolor* Ahl, 1937.
LENGTH: Males, 4¼ in. (11 cm); females, 3¼ in. (8 cm).
FIN RAY COUNTS: D, XII/9; A, XVIII/14.
DISTRIBUTION: In the original description, Ahl gives "the Netherlands Indies" (1937) as the country of origin. In those days, all the Dutch colonial possessions in that area were called the Netherlands (East) Indies (now called Indonesia). The exact locality of discovery was not known to Ahl. Even today it is not certain where in nature this species occurs. Since *M. concolor* is decidedly more sensitive to cooler water temperature than the other two *Macropodus* species, its distribution probably lies south of the other two species. There is no reliable information about its exact distribution. Schaller (personal communication) considers occurrences west or south of the Mekong River as improbable and those in Thailand and the Malay Peninsula as doubtful. The origin of this species is probably in the lowlands of Vietnam. Myers, too, gives Vietnam as its home, but he provides no more detailed distributional data. Should information about collections on the Malay Peninsula and Borneo (Sarawak) be confirmed, it will most likely be traced back to releases by humans. First importation in 1935.
DESCRIPTION: Males have strongly elongated fin tips in dorsal, anal, and tail fins. The body is compressed and moderately high, and the tail fin is deeply forked. Body coloration is a unicolored brown-gray to black-brown. A delicate net pattern formed by black-edged scales covers almost the entire body. The male becomes almost solid black during territory formation, as well as during mating and brood care, and the elongated ventral fins turn bright red. During courtship and spawning, the females often show an unclear pattern consisting of irregular diagonal stripes and a light-colored belly.
MAINTENANCE AND BREEDING: As a rule this species is well suited for a community aquarium. The males can be very aggressive toward each other, but they rarely attack other fishes. Breeding is about the same as for *M. opercularis.*

Macropodus opercularis (Linnaeus, 1758)

COMMON NAME: Macropodus or paradise fish
SYNONYMS: *Labrus opercularis* Linnaeus, 1758; *Labrus operculatus* Gmelin, 1788; *Macropodus viridi-auratus* Lacépède 1802; *Macropodus venustus* Cuvier and Valenciennes, 1846. These are only the most important syn-

onyms. The first description by Linnaeus was published in his tenth edition of *Systema Naturae* in 1758, and thus is as old as the entire system of scientific nomenclature, hence the large number of synonyms.

LENGTH: Males, to 4½ in. (11 cm); females, to 3½ in. (9 cm).

FIN RAY COUNTS: D, XIII–XVII/6–8; A, XVII–XX/11–15.

DISTRIBUTION: Korea, China, Vietnam, and the islands of Taiwan and Hainan. Lives in standing and slowly running waters of all kinds but occurs mainly in shallow bodies of water, irrigation canals, and flooded rice paddies. First imported to the West in 1869.

DESCRIPTION: Males have considerably brighter overall coloration and longer, more pointed dorsal, anal, and pelvic fins. Because of its exceptionally beautiful color and its hardiness, the paradise fish quickly became popular and has survived in the aquarium trade to the present. Because it was so easy to cultivate, later imports occurred only rarely, if at all. Even sixty years after its first importation, the grand old master of aquarology, J. P. Arnold, wrote in his book *Fremdländische Süsswasserfische* (Exotic freshwater fish) (1936), "the most beautiful, least demanding and most durable of all decorative fish." The background color is brownish red, and it has blue vertical stripes. The caudal fin and the elongated ventral or pelvic fins are red. The dorsal and anal fins are greenish blue with bluish white iridescent tips. A bright iridescent blue spot is found on the gill cover.

MAINTENANCE AND BREEDING: The males are, as a rule, excellent parents that guard their young longer than most other labyrinth fishes do. There have been numerous observations of females participating in brood care. According to an observation by Quitschau, one female continued to care for the young remaining in the aquarium after the bubblenest, a large number of the young, and the male had been removed.

This beautiful fish is not always suitable for a community aquarium. Some specimens can be quite quarrelsome, whereas others completely ignore their tankmates.

There have been many attempts to breed new strains of this popular ornamental fish. The well-known breeder of ornamental fish, Paul Matte, presented a particularly long-finned and beautiful paradise fish under the name "paradisefish" in 1893. Around the turn of the century particularly beautiful paradise fish were exhibited in so-called paradise fish shows. These fish were judged according to uniform standards. For a number of years such shows were held all over Europe. The albino form with its red vertical bars, red fins, yellowish white body, and red eyes, is the only constant color variety ever produced. Today, predominantly blue forms of this fish are showing up in the tanks of pet shops, but their appearance so far has been little more than sporadic. It is my opinion that the blue form and the albino form do not surpass in beauty the original natural coloring of this fish.

Subfamily Trichogasterinae

Genus *Colisa*

Cuvier established the genus *Colisa* in 1831. He thus removed the four threadfish of peninsular India, described by Hamilton and Buchanan in 1822, from the genus *Trichogaster* and placed them in their own genus.

The designation "threadfish" for representatives of both genera arises from their threadlike elongated pelvic fins. The species of these two genera are easily distinguished. Whereas dorsal and anal fins are about equally long-based in the *Colisa* species, the base of the dorsal fin in the *Trichogaster* species is considerably shorter than the base of the anal fin. Of the *Colisa* species of peninsular India, *C. fasciata* and *C. lalia* are well-known aquarium fish that first came to the West around the turn of the century. A third species, *C. chuna,* was imported in 1962 and also quickly found wide distribution as an aquarium fish. On the other hand, a fourth species from peninsular India, *C. sota* (Hamilton and Buchanan, 1822), has not yet been brought to the West alive. It is now believed that the specimens from which *C. sota* were described might have been females of *C. chuna.* Also up in the air is whether the aquarium fish traded under the name *C. labiosa* are a separate species. It cannot be determined at present whether these fish are identical to the *C. labiosa* described by Day in 1878 or whether Day possibly described a different fish, which to date is not known as an aquarium fish.

In its large native distribution, *C. fasciata* shows a significant amount of variation in both body shape and color. The fin ray and scale formulas of the fish sold as *C. labiosa* are within the counts for *C. fasciata*. The references to color given in the original descriptions were based on preserved specimens and are worthless. Only field investigations on living specimens with the exact data on origin could clarify this issue. Because of local politics, specimens from Burmese waters, the fish's supposed origin, cannot be obtained at this time, so this question must remain open. I worked on this problem during the 1950s, and I did succeed in obtaining live specimens of *C. fasciata* from different parts of India, but it was not possible to get live fish from Burma. The specimens I had of *C. fasciata* demonstrated quite well the great range in variation of body shape and color in this species. In crossbreeding

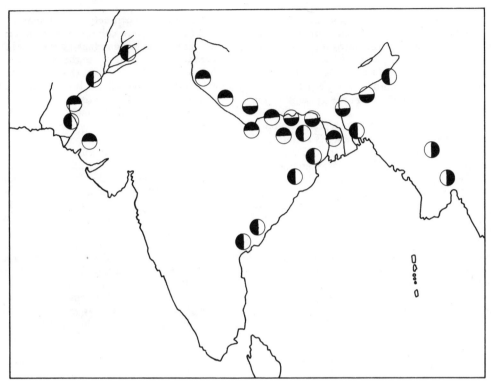

Distribution map of the genus *Colisa*

◐ *Colisa fasciata*

⬗ *Colisa lalia*

⬖ *Colisa chuna*

◗ *Colisa labiosa*

experiments between *C. fasciata* imported from India and aquarium-bred *C. labiosa,* partially fertile hybrids were obtained. From this result I concluded that the aquarium-bred *C. labiosa* is only a subspecies of *C. fasciata.* Color aside, there are some obvious differences between these fishes, particularly if one compares older specimens. Old *C. fasciata* males develop a rather wide and massive head, whereas *C. labiosa* males retain their more pointed head shape. Differences of this type are not rare in subspecies that are geographically far removed from one another, but by itself this difference does not justify a full species status. The fish sold as *C. labiosa* also differs from *C. fasciata* in temperament. *Colisa fasciata* is generally more active, requires a larger area for reproduction, and often behaves more aggressively toward the female. *Colisa labiosa* appears in the

aquarium trade only occasionally, whereas *C. fasciata* has been part of the basic aquarium-fish trade for many years. The differences between these two fish can, of course, arise in nature, but they can also come about rather quickly as a result of artificial selection in the aquarium. The description in this text of *C. labiosa* (Day, 1878) is given with the expressed understanding that it refers to the fish kept in the aquarium trade under this name. The question of whether these fish are indeed the species described by Day under that name must remain just as open as the question of the species or subspecies status of the fish. Since these fish are generally described in aquarium literature as *C. labiosa,* this name is being retained.

The natural life cycle of the *Colisa* species of peninsular India revolves around the periodic rainfall. When the rainy monsoon season begins around the beginning of July, the fish follow the rising water into the flooded lowland regions. There, in the nutrition-rich shallow water, reproduction takes place. The actual breeding season can extend to about October. When the water level recedes, the fish return to the deeper water of rivers and permanent lakes. There the growing young fish form large schools.

There are some variations in these reproduction prerequisites. *Colisa fasciata,* for example, is encountered in every imaginable kind of habitat. In certain regions of India this species is one of the most common fish; it is sold fresh or dried as a food fish for human consumption. Both Day and Regan claimed that *C. fasciata* is also encountered in northern Burma, an area that differs considerably in climate from that of the Indian lowlands. It is not known, however, under what conditions these fish reproduce there.

As aquarium residents, *Colisa* species are among the most rewarding fishes. They can all be kept in community aquariums together with smaller fishes. Specialized aquariums with several *Colisa* species can be very attractive. The smaller species, *C. lalia* and *C. chuna,* are particularly suitable for this purpose. All *Colisa* species build firm bubblenests, usually under a group of well-anchored floating plants. *Colisa lalia* adds pieces of plant matter to its nest. Intense courtship displays follow each other in rapid succession. Hiding places are needed for females when breeding is done in small aquariums, since in some cases, females are in danger of being mistaken for a male during the preliminary courtship. Females can be severely injured if they cannot escape from intensely courting males rather quickly.

Colisa chuna (Hamilton and Buchanan, 1822)
COMMON NAME: Honey gourami.
SYNONYMS: *Trichopodus chuna* Hamilton and Buchanan, 1822; *Colisa chuna* Cuvier and Valenciennes, 1831; *Trichogaster chuna* Day, 1878.
LENGTH: 1¾ in. (4.5 cm); aquarium-bred fish are often smaller.
FIN RAY COUNTS: D, XVII–XIX/7–8; A, XVII–XX/11–15.
DISTRIBUTION: Found primarily in the lowlands surrounding the lower Ganges and the Brahmaputra Rivers. Reproduction takes place mainly in the warm shallow water of the floodplains. First appeared in the aquarium trade in 1962. It is still not imported on a regular basis but is quite popular when available.
DESCRIPTION: During the breeding season the normally straw-yellow body of the male turns wine-red and the bottom half of the head, the chest, and the belly become a deep indigo blue. The dorsal fin be-

comes bright golden yellow. The female retains the normal straw-yellow color seen in the male in the nonbreeding season but develops a brown horizontal stripe that extends from the eye to the base of the tail. When the breeding season is over, the female loses the stripe and the male loses all his bright colors. During the nonbreeding period both fish look so much alike that it is difficult to tell the male from the female. The male, however, has slightly more pointed dorsal and anal fins.
MAINTENANCE AND BREEDING: This species is easy to keep in an aquarium. Because of its small size, however, it is recommended only for a community aquarium containing correspondingly small fishes. As the breeding colors develop, the male builds a thick bubblenest under anchored floating plants. If he is forced to build the bubblenest in the open at the surface, the nest is larger in diameter but consists of only one or two layers of bubbles. During courtship the male displays in front of the female, positioning himself almost vertically. During these displays the male swims toward the nest, probably attempting to lure the female toward it. After contact is made under the nest, an embrace occurs, during which ten to twenty eggs are released and fertilized. Through the entire spawning about 200 eggs are released, but nests containing larger numbers of eggs are known. The eggs hatch in twenty-four to forty-eight hours, depending on temperature. As soon as the young fish start to swim free (about two to three days after emerging from the eggs), the parents should be removed from the aquarium. The larval fish are very small but easily recognized by their dark color. The first food should be extremely small organisms, such as paramecium or rotifers.

Colisa fasciata (Bloch and Schneider, 1801)
COMMON NAME: Giant gourami, striped threadfish.
SYNONYMS: *Trichogaster fasciatus* Bloch and Schneider, 1801; *Trichopodus colisa* Hamilton and Buchanan, 1822; *Colisa vulgaris* Cuvier and Valenciennes, 1831; *Trichogaster fasciatus* var. *playfairi* Day, 1878.
LENGTH: 4 in. (10 cm); females usually a little smaller.
FIN RAY COUNTS: D, XIII–XVI/9–14; A, XV–XVIII/11–18.
DISTRIBUTION: With the exception of the extreme south and southwest, it occurs in all of peninsular India. It lives in the lowlands surrounding the big rivers, such as the Indus, Ganges, Godavari, Mahanadi, and Narmada, and in numerous smaller tributaries and other waters. Many populations are geographically isolated from each other, and this gives rise to variety in both body shape and color. First imported to the West by the fish breeder Paul Matte in 1897.
DESCRIPTION: The background color is a light olive-

brown, darker along the back and upper flanks and lighter along the lower flanks. A number of blue-green and orangish stripes run obliquely from the back to the belly. The dorsal fin is blue-green anteriorly and reddish orange posteriorly. The anal fin is blue to blue-green and has an orange outer edge. The belly and chest are blue-green, and this color extends into the anal fin. The outer edge of this fin and the threadlike pelvic fins are red. The color of the female is considerably more subdued, and the fish is smaller than the male. The oblique stripes on the sides are barely visible. A dark bar running from the eye to the base of the caudal fin is sometimes visible. The dorsal fin is pointed in the male and more rounded in the female.

MAINTENANCE AND BREEDING: The male usually builds a large bubblenest and does not use any plant matter in its construction. If floating plants are present, the bubblenest is usually built under them. *Colisa fasciata* is a very fecund species, producing up to 1000 eggs in a single spawning. At a temperature of 77 to 83°F (25 to 28°C), the eggs hatch in twenty-four to thirty hours. The young fish swim freely three days later. Early feedings should be with small food organisms, such as paramecium.

Colisa labiosa (Day, 1878)

COMMON NAME: Thick-lipped gourami.

SYNONYMS: *Trichogaster labiosus* Day, 1878; *Colisa labiosa* Myers, 1923

LENGTH: Up to 3½ in. (9 cm); females smaller; aquarium-raised specimens rarely exceed 2½ in. (6 cm).

FIN RAY COUNTS: D, XVII–XIX/7–9; A, XVII–XX/15–20.

DISTRIBUTION: According to Day, *C. labiosa* occurs in Burma, in the Irawadi River, from its mouth to the elevation of Mandalay.

A single specimen of this fish was imported to Hamburg in 1904. In 1911, a number of specimens were reported, and these fish were soon being bred quite easily. In 1936 and again in 1950, Arnold called attention to the fact that these fish did not seem to be the same species as the original single specimen imported in 1904, since they did not show any scales on the soft part of the anal fin, which the original importation did.

DESCRIPTION: Males have a brownish yellow background color and become blackish brown during the mating period. A number of faint blue-green stripes cover their sides obliquely, from top to bottom. The dorsal, caudal, and anal fins are also blackish brown. The dorsal fin has a red edge, and the anal fin has a white edge. Females are lighter in color than males. The female's dorsal fin is yellowish with a few red spots in the soft part. The caudal fin is almost colorless, and the anal fin is bluish with a rose-colored edge. The dorsal fin of the male is more pointed than that of the female.

MAINTENANCE AND BREEDING: *Colisa labiosa* is easy to keep and quite suitable for a community aquarium. Its reproductive behavior is about the same as that described for *C. fasciata*.

Colisa lalia (Hamilton and Buchanan, 1822)

COMMON NAME: Dwarf gourami, flagfish.

SYNONYMS: *Trichopodus lalius* Hamilton and Buchanan, 1822; *Colisa lalius* Cuvier and Valenciennes, 1831. In addition, fish were described under the name *Colisa unicolor* Cuvier and Valenciennes, 1831, and as *Trichogaster unicolor* Günther, 1861, both of which were probably females of *C. lalia*.

DISTRIBUTION: The most important occurrences of this species are in the areas of the Indus, Ganges, and the Brahmaputra Rivers of India. The species is found in almost all low-lying areas of northern India. It also occurs frequently in irrigation canals and rice paddies.

First arrived in the West in 1903, and it remains today the most popular and readily available import of all the *Colisa* species.

LENGTH: Up to 2¼ in. (5.5 cm).

FIN RAY COUNT: D, XV–XVII/7–10; A, XVII–XVIII/13–17.

DESCRIPTION: *Colisa lalia* is, without a doubt, one of the most colorful of all freshwater fishes. The males have an olive-brownish background color with blue-green iridescent stripes alternating with red stripes running diagonally across their sides. The fins have numerous lines and spots of the same colors. The male also has a large blue-green iridescent spot on the gill cover. The females are a drab gray, with barely visible bluish oblique stripes. The dorsal fin is pointed in the male but rounded in the female. This difference also exists in immature incompletely colored young fish, making sex differentiation a bit easier.

MAINTENANCE AND BREEDING: This species is well suited for a well-planted community aquarium in which the males will still build their bubblenests, despite minor disturbances by other fishes, and defend them tenaciously against even considerably larger fishes. *Colisa lalia* thrives best in an aquarium that provides plenty of hiding places and in which the temperature is at least 77°F (25°C). The breeding aquarium should be as large as possible. The fish should be provided with floating plants with finely divided leaves, which can be used for nest building. With the aid of small pieces of vegetation, some males build nests up to 1 in. (2.5 cm) thick. In each spawning embrace, forty to fifty eggs are released and fertilized; large females can produce as many as 600 eggs in one spawning. The eggs hatch in twenty-four hours at a temperature of 82°F (28°C), and three days later the fry begin to swim free. They are among the smallest of all labyrinth fish fry. Success in rearing *C. lalia* thus depends on whether the fry find sufficiently small food during this critical early period. An adequate supply of paramecium, rotifers, and the like should be available at this time. An additional critical period for the growing young fish occurs at about the fourth week of life. During this time the labyrinth organ comes into full development.

The quality of *C. lalia* stock currently carried in the aquarium trade is often poor; the fish are very sensitive to infection and often are carriers of diseases, such as fish tuberculosis. It is possible that this degeneration is a consequence of the selection of poor breeding stock. However, the cause may also be the indiscriminate use of antibiotics, particularly in Southeast Asia's wholesale breeding establishments. Because of the constant exposure to antibiotics, these young fish have little opportunity to develop natural immunity to bacterial diseases. Thus when the fish finally arrive from the Orient, weakened by a strenuous journey, they easily contract bacterial infections. Dwarf gourami stock should therefore be examined very carefully before purchasing.

Several true-breeding color varieties of *C. lalia* have recently become available. All-blue and all-red forms began to appear in the trade around 1980. At first, only males were sent from the Orient, where they originated. Later, females appeared in shipments with males, but they turned out to be females of the normal wild type. Thus breeding them resulted in many throwbacks to original wild-type coloration. Individuals of this early red form lost their color after a time, gradually becoming similar to the wild form but lacking their expected color intensity. The red color of these specimens may have been derived from their food. Of all the new color strains developed so far, only one, a copper-red form with an extremely beautiful blue dorsal fin, has proved hereditarily stable. In the fish trade this form became known as the rainbow lalia. The females of this color form are a light gray color without any striped markings. When a copper-red male is mated with the corresponding female, the offspring have the same color as the copper-red male. In mating with wild-type *C. lalia*, the corresponding heredity factor is passed on recessively. In my breeding work, a mating between a copper-colored male and a natural-colored female resulted in the F_1 generation having exclusively natural-colored offspring. However, in further matings of the offspring with each other, red males again appeared in the F_2 generation.

Genus *Trichogaster*

The genus *Trichogaster*, established by Bloch in 1801, contains four valid species, and one of them, *T. trichopterus*, is divided into two different subspecies. *Trichogaster* species show similarities in appearance to *Colisa* species. For example, as in *Colisa* species, the pelvic fins are elongated into threadlike appendages, which are used in sensing the environment as well as in part of the mating ritual.

Their range lies east of the range of *Colisa* species (India and Burma). *Trichogaster* species are found in Vietnam, Kampuchea, Thailand, Malaysia, and to the southeast, in Sumatra, Borneo, Java, and Bali. Occurrences east of the zoogeographic borderline known as the Wallace Line (i.e., on the small Sunda islands of Lombok and Sumbawa, on the island of Obi [part of the Molluccas], and on the Philippine island of Luzon) probably came about through accidental release by fish hobbyists or intentional releases by scientists and governments. This theoretical zoogeographic border, named by the British zoologist Alfred Russell Wallace, separates the animal world of the Oriental region from the animal world of the Australian region in the Indo-Malayan area with surprising accuracy. Thus Borneo, for example, possesses a freshwater fauna different from that of neighboring Celebes. This border divides the islands of the Malay Archipelago into two parts and runs between the islands of Bali and Lombok and between Borneo and Celebes. Continuing north, it runs west of the Philippines. It can be assumed that occurrences of labyrinth fishes east of this line are very recent, with humans usually responsible for their presence.

The largest *Trichogaster* species, the snakeskin gourami *(S. pectoralis)*, which grows to a length of over 8 in. (20 cm), was released as a human food fish in waters far removed from its original distribution. Today this species not only occurs on Sri Lanka but is also encountered on several islands of the Antilles in Central America.

Trichogaster species are much larger than *Colisa* species, all four exceeding 4 in. (10 cm) as adults. The largest, *Trichogaster pectoralis*, reaches its adult size of 8 to 10 in. (20 to 25 cm) only in very large aquariums. All *Trichogaster* species are sexually mature at about half of their adult size. Juvenile specimens are easily distinguished from *Colisa* species by the considerably shorter length of the dorsal fin base. The threadlike pelvic fins are also shaped somewhat differently. In *Colisa* species they consist of a single threadlike elongated fin ray, but in *Trichogaster* species, close inspection reveals that the long filamentous fin is composed of two fused fin rays. The caudal fin is always slightly

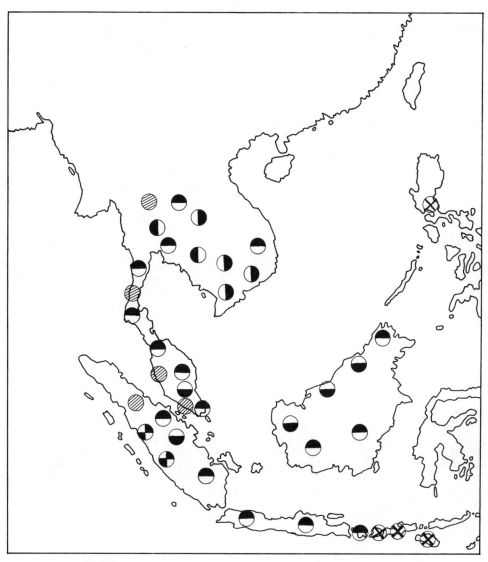

Distribution map of the *Trichogaster* species

- Trichogaster leeri
- Trichogaster microlepis
- Trichogaster pectoralis
- Trichogaster pectoralis (transplanted populations)
- Trichogaster trichopterus trichopterus
- Trichogaster trichopterus sumatranus
- Trichogaster trichopterus trichopterus (populations probably transplanted)

more forked in *Trichogaster* species. In sexually mature fishes or specimens nearing sexual maturity, the sexes of *Trichogaster* species are easily distinguished; the dorsal fin of the male is considerably longer and more pointed than that of the female. The color differences between males and females are much less distinguishable in *Trichogaster* species than they are in *Colisa* species.

Even though they grow quite large (for an aquarium fish), *Trichogaster* species show a preference for small food organisms, such as adult brine shrimp, mosquito larvae, or tubifex worms, over larger organisms such as earthworms. This is probably related to the fact that their mouths are not very deeply cleft for fishes of their size. These fishes also readily consume vegetable matter, such as algae, and, if presented, will also eat bits of boiled spinach. They also readily accept dry flake foods, but like most fishes that eat these prepared foods, they grow better, remain healthier, and live longer on a mixed dry and live or fresh food diet than they do on a dry food diet alone. In fact, if well nourished, with plenty of roughage included in the diet, it is not at all unusual for these fishes to live for seven or eight years or even longer.

In slowly flowing waters *Trichogaster* males seek out sheltered areas in which to build their bubblenests. In these areas the nests are often found between the floating leaves of water lilies, where the surface currents are practically nonexistent. In standing waters, areas covered by floating plants are preferred for bubblenest building. During bubblenest building the males breathe in air at the surface and release it through the mouth and the gill cover openings as a spray of fine mucusencapsulated bubbles. In an aquarium these fishes usually build their bubblenests solely of bubbles, rarely using pieces of vegetation, as their close relatives, the *Colisa* species do. The exception is the moonbeam gourami, *Trichogaster microlepis,* which, in the aquarium as well as in the wild, often strengthens its nest with small pieces of vegetation.

During courtship the male and female remain rather close to each other, rather than the male "leading" the female to the nest, as is seen in most other bubblenestbuilding species. In fact, in these fishes most matings are initiated by the female, which spontaneously seeks out the male under his bubblenest. After a few brief displays by the male, the female nudges the male in the side with her snout. A nuptial embrace usually follows, during which the male arches himself under the female, with their genital openings in close apposition. After twenty to thirty seconds, and holding this position, the pair rotates in such a way as to bring the female upside down. In this position, and after several false starts, the embraced pair quivers, and a shower of eggs and sperm is ejected upward toward the nest. Forty to eighty floating eggs are released in each embrace. Large, fully mature females may release a total of 1500 to 2000 eggs.

At a temperature of 75 to 77°F (24 to 25°C) the eggs hatch in about thirty-six hours. Hatching can be accelerated to twenty-four hours by raising the temperature up into the range of 82 to 85°F (28 to 30°C). Three days later the larval fish begin to swim free and feed. While the fry grow quite rapidly, they generally do not grow at a uniform rate, and after a few weeks great size disparities develop among fry. In order to encourage growth in the smaller fry, so that the entire brood may be raised successfully, it is necessary to sort the fry by size and separate them into several groups, keeping each group in a separate aquarium.

Trichogaster species are by and large peaceful fishes. They do not scrap excessively among their own kind, and they get along fairly well with most other fishes that are about the same size. Before they reach sexual maturity, they even get along well with other fishes much smaller than themselves. Once sexual maturity is achieved, however, only the pearl gourami, *Trichogaster leeri,* can be trusted almost completely in the presence of very small fishes. After that point the others have a tendency to pick on smaller fishes, especially if the aquarium is crowded and not well planted. On the whole, however, *Trichogaster* species are a fairly docile group of fishes that make good residents for a community aquarium, especially one containing an assortment of these species.

Trichogaster leeri (Bleeker, 1852)
COMMON NAME: pearl gourami, pearl leeri
SYNONYMS: *Trichopus leeri* Bleeker, 1852; *Osphronemus trichopterus* var. *leeri* Günther, 1861; *Trichopodus leeri* Regan, 1909.
LENGTH: Up to 4¾ in. (12 cm).
FIN RAY COUNTS: D, V–VII/8–10; A, XII–XIV/25–30.
DISTRIBUTION: Malay Peninsula, Sumatra, and Borneo. Its presence in Java is doubted, and its occurrence around Bangkok is probably the result of an accidental escape from a fish breeder, or, as assumed by some, a case of mistaken identity. This species was first imported in the early 1930s.
DESCRIPTION: The body is rather broad from top to bottom and is laterally compressed. The background color is silvery gray. In front of the dorsal fin the back is an olive-brown color. Numerous small light spots, each with a dark edging, cover most of the body as well as the anal, dorsal, and caudal fins. The mosaic pattern formed by these spots appears almost iridescent. The markings are nearly identical in both sexes but more intense in the male. Full-grown males also

have a significantly longer and more pointed dorsal fin and a much larger anal fin with longer extended fin rays, which gives the fin the appearance of having an irregular-looking lacy fringe. The most beautiful adornment of the males is the reddish orange color of the underside, which extends from the lower part of the head, over the chest, to the front half of the abdomen and includes the front portion of the anal fin. During the mating period and during brood care the reddish orange color takes on a spectacular intensity.

MAINTENANCE AND BREEDING: This species usually builds large bubblenests. Released eggs are supported by or imbedded in the bubbly foam of the nest. Plant pieces are rarely used by this species for nest building. The species is quite fecund, producing up to 1000 eggs or more in one spawning. Raising a large brood requires an abundance of small first foods such as paramecium. Brine shrimp nauplii are accepted from about the sixth day after the young become free swimming.

Trichogaster microlepis (Günther, 1861)
COMMON NAME: Moonbeam gourami.
SYNONYMS: *Osphronemus microlepis* Günther, 1861; *Trichopus microlepis* Sauvage, 1881; *Trichopodus microlepis* Regan, 1909.
SIZE: To 6 in. (15 cm).
FIN RAY COUNT: D, III–IV/7–10; A, X–XI/34–40.
DISTRIBUTION: Thailand and Cambodia. First brought to the Western aquarium trade in 1951.
DESCRIPTION: The basic color is a satiny monochromatic silver, which in top or front lighting has an iridescent blue shimmer. In adults the sex differences are not as distinct as they are in the other *Trichogaster* species, but they are nonetheless distinguishable. The threadlike pelvic fins vary from orange to red in the male and are yellowish in the female. Also, the dorsal fin of the male is longer and slightly more pointed than that of the female. Mature females are usually more robust than males.

MAINTENANCE AND BREEDING: The moonbeam gourami differs from the other *Trichogaster* species in how the male builds his nest. If there are enough plants available, especially floating plants, such as water sprite (*Ceratopteris thalictroides*), the male gathers and assembles large mats of this material. He may also tear off and use pieces from rooted plants. Mucus-encapsulated bubbles are used to hold the mat together. If suitable plants are not available, the male builds a normal bubblenest. I have never observed them not to use plants, if they were available. Nests consisting of large piles of rapidly decomposing plants provide an excellent reproductive and growth medium for various kinds of microorganisms, which serve as a choice first food for the tiny fry during their first few days of life. In large heavily-planted aquariums these fish can build a nest as large as 12 in. (30 cm) in diameter with a center height of up to 6 in. (15 cm).

Trichogaster pectoralis (Regan, 1910)
COMMON NAME: Snakeskin gourami.
SYNONYMS: *Trichopodus pectoralis* Regan, 1909; *Trichogaster pectoralis* Smith, 1933.
LENGTH: Up to 8 in. (20 cm); one local form is even larger.
FIN RAY COUNT: D, VII/10–11; A, IX–XI/36–38.
DISTRIBUTION: Originally native to South Vietnam, Kampuchea, and eastern Thailand, this species is now also found in many other Southeast Asian localities. As early as 1920 it was established around Singapore, and a few years later it was released in other parts of the Malay Peninsula. Today, in many areas of the Malay Peninsula this is one of the most commonly occurring freshwater fish species. Other released populations are found on Sri Lanka, on two islands of the Antilles, and also on Haiti. The snakeskin gourami is found in almost all types of waters; in rivers and canals, in ponds and irrigation ditches, in flooded rice paddies, and even in the brackish water of the Mekong Delta.

This species was first imported as an aquarium fish in about 1914, but only in small numbers. Larger importations began in about 1952.
DESCRIPTION: Greenish gray with silver sides. Along the center of the body there is a horizontal stripe that consists of a number of dark spots. Between these spots, from the back to the belly, are lighter stripes. The horizontal stripe disappears completely during the spawning period and returns when spawning and brood care is complete. It also fades away as the fish ages.

The male is somewhat more slender than the female, although both sexes have more girth than all other *Trichogaster* species. The male has a longer more pointed dorsal fin than the female. According to the well-known aquarist H.J. Richter, this species builds a relatively small bubblenest. Full-grown females lay 2000 to 3000 or more floating eggs. Because of the large size of this fish and the large number of eggs it produces, only a very large aquarium is recommended for its breeding.

Trichogaster trichopterus (Pallas, 1777)
COMMON NAME: Blue gourami, three-spot gourami.
SYNONYMS: *Labrus trichopterus* Pallas, 1777; *Trichogaster trichopterus* Bloch and Schneider, 1801; *Trichopodus trichopterus* Lacépède, 1801; *Trichopus trichopterus* Cuvier and Valenciennes 1831; *Trichopus sepat* Bleeker, 1845; *Osphromenus siamensis* Günther, 1861; *Trichopus siamensis* Sauvage, 1881; *Osphronemus saigonensis* Borodin, 1930.

Descriptions of Species

The numerous synonyms can be explained in part by the fact that many local forms deviating somewhat from each other were originally described as species.

LENGTH: 5 in. (12 cm)

FIN RAY COUNT: D, VII–IX/8–10; A, X–XII/33–38.

DISTRIBUTION: The species is very common throughout Indochina and the Malayan area (see map). Occurrences in the marginal areas of its distribution came about by releases. This species is quite variable over its large geographical range, and there are a large number of local forms. A detailed taxonomic study of these local forms has not yet been undertaken. Up to now only a mottled blue form from Sumatra has been described as a subspecies. This fish occurs in almost all types of waters but is very common in overgrown river banks, canals, ponds, and lakes in which there is an abundance of vegetation. It also lives periodically in flooded rice paddies and is even temporarily found in brackish water.

DESCRIPTION: Typical coloration consists of two dark spots, one of which is in the center of the side, and the other is located on the caudal peduncle. The common name, three-spot gourami, is derived from these spots plus the dark eye, which is considered the third "spot." The basic color and the markings can be quite different, depending on the place of origin of the specimens. Some individuals have a beautifully marbled pattern on a bluish gray or greenish gray background. Others are monochromatically bluish gray, but there is also a strain that has a more brownish background color.

The dorsal fin in the male is considerably longer and more pointed than that of the female. The anal fin is bigger on the male and marked on the outer edge by rows of iridescent orange or yellow dots.

MAINTENANCE AND BREEDING: Of all the forms of this genus, this one seems to be the hardiest. It can tolerate slightly lower temperatures than the others or the other *Trichogaster* species. While it usually breeds in a manner typical for the genus, it has been known to breed using only a scant scattering of bubbles that barely resemble a nest at all.

Trichogaster trichopterus sumatranus Ladiges, 1933.

COMMON NAME: Blue gourami.

DESCRIPTION: A subspecies with a powder blue background color; as far as can be determined, this is a wild color mutation. First imported in 1933, it quickly became popular as an aquarium fish. Today it is kept more commonly than most other forms of the species. There are hardly any marks at all on this fish, except for the two body spots, but even these look more or less washed out in adult specimens.

BREEDING FORMS of *trichogaster trichopterus:*

Several specially bred color forms of *T. trichopterus* have come into the aquarium trade during the past few decades. In most cases it is not possible to determine whether a particular color form was bred out of the original described form or from the blue subspecies *T. t. sumatranus.*

These forms should just be given common names and no attempts to taxonomically classify them, as has unfortunately been done in much of the aquarium literature, should be attempted. Designations such as var. (= variety) or forma (= form) should not be used for such breeding forms, since their origins or derivations are really unknown. The best known of these breeding forms are the gold gourami and the silver gourami.

GOLD GOURAMI

The gold gourami has a golden yellow background color with either black or reddish brown eyes. Most specimens show a clearly conspicuous marble pattern. It is usually assumed that this breeding form was developed from a xanthistic (yellow-colored) mutation of the originally described form.

SILVER GOURAMI

The silver gourami has a silver background color. There are some individuals that have a clear marbled pattern and others in which this pattern is only weakly apparent. Nothing is known about the origin of the first specimens. In contrast to the gold gourami, the silver gourami has not had much distribution in the aquarium trade.

COSBY, OR MARBLED, GOURAMI

The Cosby gourami, or marbled gourami, was probably developed in the United States. The original specimens seem to have been *Trichogaster trichopterus sumatranus.* When they first color up the Cosby gouramis have a few dark blue spots on a powdery light blue background. As they grow, however, these spots are replaced by a dark marbled pattern. These strikingly marked fish are very popular in the aquarium hobby.

TOP: The snakeskin gourami, *Trichogaster pectoralis,* originally native to Cambodia (now Kampuchea), eastern Thailand, and southern Vietnam, is a valued food fish in many areas of Southeast Asia.

BOTTOM LEFT: The gold gourami is an aquarium bred color variety of the blue gourami, *Trichogaster trichopterus.*

BOTTOM RIGHT: A dark color phase of the three-spot, or blue, gourami, *Trichogaster trichopterus trichopterus.*

This color form has been designated with all kinds of pseudo-scientific names, such as *"Trichogaster cosby,"* *"Trichogaster sumatranus,* var. *cosby,"* or *"Trichogaster sumatranus* forma *cosby."*

Subfamily Sphaerichthyinae

Genus *Parasphaerichthys*

Parasphaerichthys ocellatus Prashad and Muckerji, 1929

This monotypic genus closely resembles the genus *Sphaerichthys* and was established for this species only. The species was originally described from two half-grown specimens about 1 in. (2.5 cm) long, which originated in northern Burma. In the early 1980s a few living specimens of this extremely rare species were finally imported alive. Their appearance is best described by the photo on page 88.

Genus *Sphaerichthys*

This genus was established by Johann Canestrini in 1860. As far as can be determined, its description was based on a single specimen, the exact origin of which was unknown; the statement by Canestrini that India was its place of discovery is not correct. Based on later studies by others it is agreed that even the fin ray count on Canestrini's original specimen was wrong. These faulty descriptions caused former Hamburg Zoological Museum Director George Dunker to undertake a new description of the chocolate gourami which was published under the name *Osphromenus malayanus* Dunker 1904. This description, although also incorrect because of faulty taxonomic procedure, is much more exact than the one by Canestrini, and it is based on material known to be from the Malay Peninsula.

TOP LEFT: The three-spot, or blue, gourami, *Trichogaster tichopterus,* is the color phase most commonly seen in the aquarium trade.
TOP RIGHT: The Cosby gourami, *Trichogaster trichopterus sumatranus,* is a subspecies of the blue gourami.
CENTER LEFT: *Parasphaerichthys ocellatus,* a rarely seen species that is only occasionally imported singly from northern Burma.
CENTER RIGHT: A male of the pointed-head chocolate gourami, *Sphaerichthys acrostoma.*
BOTTOM LEFT: *Sphaerichthys osphromenoides selatanensis,* a possible subspecies of the chocolate gourami.
BOTTOM RIGHT: The chocolate gourami, *Sphaerichthys osphromenoides.* This species was captured on the Malay Peninsula.

Only the species *S. osphromenoides* was known before the year 1930. In that year, Pellegrin described an additional species from Borneo under the name *S. vaillanti.* It remains an open question whether this species, which apparently never reached the western hemisphere alive, is indeed a separate species, or whether it is only a subspecies of *S. osphromenoides* occurring on Borneo, as assumed by J. Vierke. With the proof of a *Sphaerichthys* on Borneo, however, the area of distribution of this genus was considerably expanded. E. Korthaus and W. Foersch, on a 1978 collecting trip to South Borneo, brought back live specimens of another subspecies of *S. osphromenoides* as well as a species of this genus that was previously unknown, thus establishing that the genus was in fact distributed on the island of Borneo, but how widely distributed still remains unknown.

Even the distribution of *S. osphromenoides,* a species which has been known for quite some time, is incomplete. It may be mentioned in this connection that the two islands of Sumatra and Borneo cover an area which is five times as large as the state of Oregon. Many of the waters on these islands are still unexplored, especially on Borneo, where periodically there are broad flooded areas.

These fishes mainly seem to live in shallow waters of all kinds: in swampy areas heavily overgrown with plants but just as often in acidic water holes lacking any plant growth; in slow-flowing, frequently polluted canals; and sometimes in coffee colored brooks of tropical rain forests. Most of these waters are characterized by an extremely low mineral content. Some of the water conditions in which they were found were so extreme that the fish had to be living mainly or at least temporarily, on insects landing on the surface, because the water was too acidic for the survival of small aquatic organisms. Based on the nature of these environments, it is probable that the distribution of the chocolate gourami frequently consists of a number of populations separated from one another.

The literature on the reproduction and behavior of *S. osphromenoides,* the only member of the genus thus far bred in the aquarium, is fraught with contradictions. Reichelt, who first brought live chocolate gouramis out of Southeast Asia alive in 1905, first assumed that the species was live-bearing. He had found "newborn" young fish in the transport water during a repacking in Singapore. Later it was found that members of this species are mouthbrooders, the female being the brood-caring member. When this species was imported more regularly, beginning in the middle 1950s, reports appeared in which these fish were supposed to be bubblenest builders. Other reports mentioned a symbolic bubble spitting as an introduction to court-

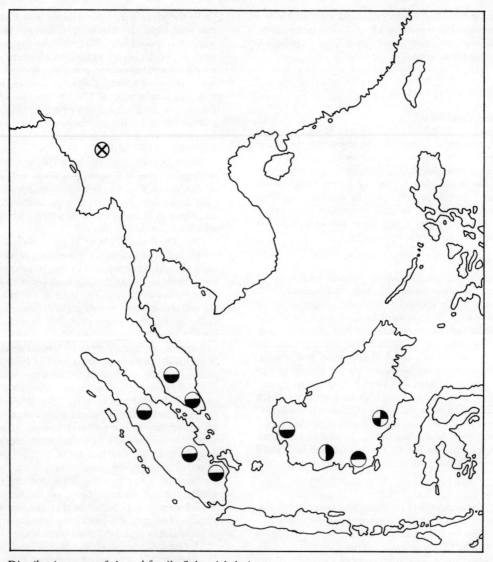

Distribution map of the subfamily Sphaerichthyinae

⊗' *Parasphaerichthys ocellatus*

◖ *Sphaerichthys acrostoma*

◕ *Sphaerichthys osphromenoides*

◒ *Sphaerichthys osphromenoides selantanensis*

◐ *Sphaerichthys vaillanti*

ship behavior and to spawnings in which the eggs were later taken into the mouth from rudimentary bubblenests. Several years later, clear-cut mouthbrooding was observed, and also photographed, in which the female acted as the mouthbrooder. Many aquarists thereafter ignored the fact that the older statements were incorrect. From my observations, I agree that *S. osphromenoides* is indeed a mouthbrooder.

Information on the mouthbrooding care of *S. osphromenoides,* and on mouthbrooding labyrinth fishes as a whole provides evidence that these brood-care forms are a recent evolutionary development. Certain aspects of the brood-care behavior appear to be not well established, and broodcare is frequently broken off prematurely. In the early 1950s I observed in imported fish of *S. osphromenoides* both foam spitting as well as delayed egg-gathering from rudimentary bubblenests, but I was not able to observe the actual deposition of the eggs. The eggs were taken into the fishes' mouths when they fell from the few foam bubbles the fish had produced. In the mating behavior of fish from more recent imports there are reports that a partial embrace of the pair near the bottom of the aquarium, without the female being turned over. The released eggs were picked up directly from the bottom. They had a diameter of 1.2 to 1.5 mm and a light yellowish hue. So far I have never been able to observe a female succeed in picking up all released eggs. Some eggs, for which there was obviously no more room in the mouth, were always left over. In most spawnings the eggs disappeared from the mouth three or four days later, but it could not be determined whether they had been eaten or spit out. Similar observations have also been made by other aquarists (F. Schneider and D. W. Godfrey). To this day it cannot be claimed that systematic reproduction of this species in an aquarium has been successful. If young fish are hatched they are released from the mouth after thirteen to fourteen days. At that time they are 5 to 6 mm long and, thanks to their dark pigmentation, are easy to find. The young mostly stay close to the surface between floating plants. They can be fed immediately with newly hatched brine shrimp nauplii. Since this food quickly sinks to the bottom, it is advisable to transfer the young fish carefully, with a spoon, so as to avoid exposing their delicate gills and developing labyrinth organs to the atmosphere, to shallow glass bowls that have been filled with water from the aquarium in which the fish were bred.

Chocolate gouramis do best in soft, and somewhat acidic water. They are sensitive to high nitrate concentrations as well as to skin parasites and to an accumulation of microorganisms in the water. It is therefore recommended that they be kept in a special aquarium in which these stringent environmental factors can be regulated. Kept by themselves, chocolate gouramis are very shy and hide between the plants at the slightest disturbance. Association with a few small fish which are not so shy often helps them to overcome their own shyness. These fish should also be able to survive in this acidic and sterile environment. *Rasbora heteromorpha* make excellent tankmates for them. Larger and heavily planted aquariums are fine. The water should be as soft as possible, with the hardness not exceeding 10 DH. The water should be filtered over peat moss, which softens and acidifies it. Particular attention must be given to feeding. Of all labyrinth fishes, the chocolate gourami is probably most susceptible to organ adiposis (fat deposits in vital organs). The best foods for these fish are small mosquito larvae, cyclops, and smaller daphnia. Dry food is sometimes accepted by some individuals. Chocolate gouramis seem fond of small surface-dwelling wingless fruitflies.

When frightened, chocolate gouramis take on a light coloration; this is also true for the other *Sphaerichthys* species. At times, they can be quite quarrelsome among themselves, but the skirmishes rarely take on damaging proportions. They are peaceful toward other fishes, yet association with most other fishes is not recommended, the main reason being the special conditions they require.

Sphaerichthys acrostoma Vierke, 1979
COMMON NAME: Pointed mouth chocolate gourami.
DISTRIBUTION: South Borneo. The species was discovered in 1978 by E. Korthaus and W. Foersch about 250 km northwest of Banjarmasin.
LENGTH: About 2¾ in. (7 cm).
FIN RAY COUNT: D, VI/9; A, IX/20.
DESCRIPTION: This species differs from other species of chocolate gouramis by its larger size and more pointed snout. According to W. Foersch, mature males are light gray and have a distinctly swollen-looking throat area. The females appear a little more stocky and the throat area is not as swollen-looking, which makes the head appear more pointed than that of the male. This is a more active species than *S. osphromenoides*. Several times females have been observed with eggs in their mouths, but after a few days the eggs mysteriously disappeared.

Sphaerichthys osphromenoides Canestrini, 1860
COMMON NAME: Chocolate gourami
SYNONYM: *Osphromenus malayanus* Dunker, 1904
LENGTH: Up to 2¼ in. (5.5 cm).
FIN RAY COUNT: D, VII–XII/7–10; A, VIII–X/18–22.

Information on the spines and soft fin rays is given with reservation. The counts given are based on an average of the number of rays so far ascertained. Thus they include the high values given by Canestrini. Almost all chocolate gouramis have fewer spines in the dorsal and anal fins than Canestrini states. This is particularly true of fishes from Borneo, for which Vierke gives the fin ray count D, VII/8–9; A, VII/21–23.
DISTRIBUTION: Malay Peninsula, Sumatra, Borneo. This species occurs predominantly in mineral-poor water with hardness values between 0.5 to 5.0 DH and with pH values of 5.0 to 6.5. This species was first imported in 1905. It was imported again in 1934. It has been imported semi-regularly since 1950 and is occasionally available in the aquarium trade.
DESCRIPTION: The body shape is oval and compressed laterally. The snout is tapered and the eyes are relatively large. Fish in good health show a deep, rich brown coloration. A small stripe of yellowish white runs from the forehead to the eye. A wider stripe of the same color runs from the neck over the gill cover to the base of the pelvic fins. Four more stripes appear on the side of the body. The first one is approximately over the center of the body, running almost vertical. The second runs from the posterior end of the base of the dorsal fin to about the halfway point along the base of the soft part of the anal fin. The third stripe runs across the anterior caudal peduncle, and the fourth runs right along the posterior caudal peduncle. The third stripe can be completely missing in some specimens or sometimes only the upper half is visible.

The dorsal fin is pointed in the male, a characteristic not always visible, because the dorsal fin is spread out only rarely. In the female, the dorsal fin is rounded.

During the mating season ripe males have lightly edged fins, and ripe females become visibly round. Not always visible in females is a faint horizontal stripe along the midline of the body.

Sphaerichthys osphromenoides selatanensis Vierke, 1979
A subspecies of the chocolate gourami from southeast Borneo. This subspecies differs from *S. osphromenoides* in part by its smaller number of spines both in the dorsal and anal fins and by significant differences in the color pattern. When in good condition *S. o. selatanensis* shows a clearly visible horizontal stripe and five diagonal stripes. The rear part of the ventral edge of the body is decorated with a yellowish white border.

Sphaerichthys vaillanti Pellegrin, 1930
This species was described by Vaillant in 1893, but he did not recognize its status as a species and confused it with *Ctenops nobilis*.

DISTRIBUTION: Southern Borneo. Specimens have been exported to the West only rarely and only in the last few years.
LENGTH: Up to 2 in. (5 cm).
FIN RAY COUNT: D, VII–VIII/7–8; A, IX–XI/16–18.
DESCRIPTION: The background color is copper-red. About 10 light diagonal stripes run over the body (according to Ladiges). It is uncertain whether this species is truly a separate species or a subspecies of *S. osphromenoides*.

Subfamily Ctenopinae

Genus *Betta*
Species of the genus *Betta* occur from Thailand, Kampuchea, and South Vietnam all the way to Java in the south. The taxonomy of bettas is difficult and often extremely confusing. One reason for this is that some of the species have been redescribed so often that there are numerous synonyms. Another reason is that species scientifically unknown until now are being discovered and not only in the little explored waters of Borneo.

All *Betta* species lay sinking eggs. In this genus there are both bubblenest builders and mouthbrooders. Bubblenest-building species can be distinguished from mouthbrooders by their appearance. Mouthbrooding species, as an adaptation to their brood-care behavior, have larger heads than the bubblenest-building species, particularly in the males, which are the ones that look after the brood.

Bubblenest-building *Betta* species
Of the known species of *Betta,* six are bubblenest builders and one other species probably is a bubblenest builder. These species occur in Laos, Kampuchea, Thailand, the Malay Peninsula, Sumatra, and probably also in

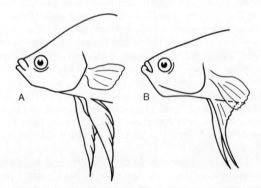

Head forms in the genus *Betta*
A = bubblenest builder type   B = mouthbrooder type

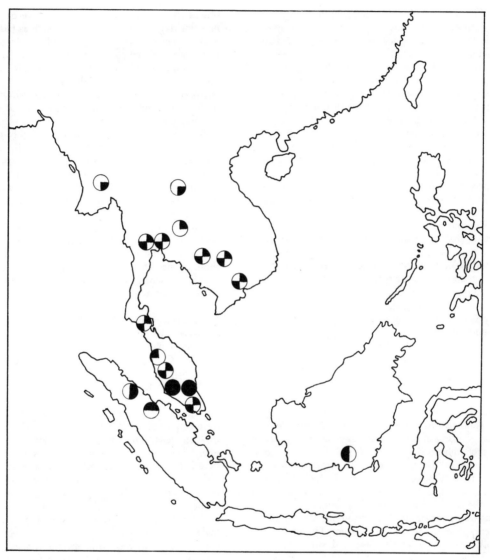

Distribution map of bubblenest-building *Betta* species

◐ *Betta bellica*   ● *Betta imbellis*

◒ *Betta coccina*   ◑ *Betta smaragdina*

◑ *Betta fasciata*   ◕ *Betta splendens*

◑ *Betta foerschi*   ◔ *Betta splendens* (probably
                          populations resulting from releases)

Borneo. *Betta splendens* is now found in numerous areas in which it originally was not native. It is no longer possible to outline precisely what its original distributional area was.

In nature, the bubblenest building *Betta* species are frequently found in overgrown waters of lowlands, in ditches, ponds, and similar standing waters. Where they occur in running water, they can move into temporarily flooded areas. The males build solid nests of relatively large air bubbles (compared to those of the gouramis). During nest construction they normally spit out one bubble at a time, not several bubbles simultaneously, as other labyrinth fishes do. The nest is frequently attached to plant stalks or floating leaves. Only the truly ripe female is tolerated in the vicinity of the nest or even in the same aquarium with the male. The females show their readiness to spawn by not fleeing from the male when he displays by spreading his fins. *Betta splendens* females also show their readiness by undergoing a marked color change to a dark, mottled or irregular pattern of wide vertical bars. After the male assumes his display pose, a ripe female swims toward him, and she assumes an oblique, head-down position. Spawning is usually preceded by several pseudo-spawnings in which no eggs are released. Once the fish have started spawning, the pair rarely goes far from the nest; however, sometimes the female may flee the area if the male gets too rough. The female is turned around during the embrace so that her belly is upward. It is not uncommon, however, that the turn is only partly executed. The released eggs have a yellowish to white color and can easily be recognized. During mating, the eggs, after they have been expelled from the genital papilla, first glide along the anal fin of the male and then drop toward the bottom. Usually before the eggs reach the bottom the male releases himself from the embrace and starts to gather them with his mouth. The female frequently remains in an upside down position and drifts, seemingly enraptured, toward the bottom. The male spits the collected eggs into the nest and supports them with additional air bubbles before another pairing occurs.

In *B. splendens* the number of eggs in one spawning can be 200 to 300 and in exceptional cases even up to 500. In females of *B. imbellis* and *B. smaragdina,* on the other hand, only 100 to 150 eggs are normally produced. According to investigations by Stanislaus Frank, the eggs of *B. splendens* are smaller than those of *B. imbellis* (0.8 mm diameter and 1.05 mm diameter respectively). According to Frank's studies, the young fish of *B. imbellis* develop faster than those of *B. splendens.* Although the young of bubblenest-building bettas can accept newly hatched brine shrimp nau-

plii as their first food, they may thrive better for the first few days on microorganisms such as paramecium.

Betta imbellis, B. splendens, and *B. smaragdina* are very closely related to one another and can be crossbred. As far as I know, no further spawnings have been accomplished with individuals from hybrids of *B. splendens* and *B. smaragdina.* But according to H. J. Richter, the hybrids of *B. splendens* and *B. imbellis* do reproduce with each other.

Betta bellica and *B. fasciata* are probably genetically very closely related, if they aren't actually geographic subspecies of the same species. Unfortunately, these fishes are rarely brought to the West. When they are it is only as a few individual specimens, so our information, so far, allows no conclusions about their behavior.

Due to the extremely problematical taxonomy of the *Betta* species, in the following species descriptions in addition to the counts of the dorsal and anal fin rays, those of the other fins are given, as is the scale count. A correct identification of a species may not always be possible even with this information.

Betta bellica Sauvage, 1884
LENGTH: 4½ in. (11 cm).
FIN RAY AND SCALE COUNTS: D, I/9–10; A, II/27–30; Pel. I/5; C/11; L.l. scales, 35; Scale Rows, 9½.
DISTRIBUTION: Malay Peninsula and the lowland area of the Perak River and its tributaries. First importation to the West occurred in 1905. Later only a few individual specimens, and almost always only males, were imported.
DESCRIPTION: The background color of males is yellowish brown to reddish brown. The scales on the sides have bright green iridescent edges. The fins have reddish brown rays with green iridescent soft tissue in between the rays. In ailing specimens 4 to 5 diagonal stripes appear. No details are known about the color of the female. Presumably it is similar to that of the male, only perhaps not as intense. Although not as feisty as *B. splendens,* this species is said to be used sometimes for staged fish fights in Malaysia.

A Swedish aquarium enthusiast with whom I am acquainted brought back two males of this species from Malaysia, where he caught them in a swamp hole at the Kinta River, a tributary of the Perak River. The males built large bubblenests and courted females of *B. splendens.* No spawning ever took place, however. In the aquarium these fish should be cared for in the same manner as *B. splendens.*

Betta coccina Vierke, 1979
LENGTH: 1¾ in. (4.5 cm).
FIN RAY AND SCALE COUNT: D, II/7–10; A II,/24–26,

Pel. I/5, L.l. scales, 29 + 2–30 + 2.

DISTRIBUTION: Sumatra. First importation into the United States came in the mid 1980s in a shipment of other fishes from Jakarta.

DESCRIPTION: When half grown, males show a dark wine-red background color. The eyes and a spot on the side, located below the anterior base of the dorsal fin, are bright green. Full-grown males have a more brownish background color. The iridescent sidespot fades with advancing age. According to Schaller, this species is a bubblenest builder.

Betta fasciata Regan, 1909
LENGTH: 4 in. (10 cm).
FIN RAY AND SCALE COUNT: D, I/9–10; A, II/30, L.l. scales, 34–36.
DISTRIBUTION: Sumatra. In small bodies of water that frequently are heavily polluted. This fish was first imported in 1906 under the name *Betta bellica*; later only a few specimens were imported.
DESCRIPTION: This is a slender cylindrical-shaped fish. The basic color is dark blue. Scales on the side are decorated with small colored spots. Nine to ten dark diagonal stripes are shown by this fish but usually only when under stress. Judging from its appearance, especially the small head in relationship to body length, this species is a bubblenest builder.

Betta foerschi Vierke, 1979
LENGTH: 2½ in. (6.5 cm).
FIN RAY AND SCALE COUNTS: D, O–I/8–9; A, I/5; L.l. scales, 31 + 2–4.
DISTRIBUTION: The fish were first caught in the river system of the Mentaya 150 mi (250 km) northwest of Banjarmasin in South Borneo.
DESCRIPTION: The body is long, thin, and laterally compressed. The head is small and wide. An elongated soft anal fin ray reaches quite a bit beyond the posterior base of the anal fin when this fin is laid back. The background color is almost black, the dark color also occurring on the fins. Two vertical stripes on the gill covers are gold-colored in the male and red in the female. When frightened, these fish show a number of lighter vertical stripes, while the back remains a light color. The vertical stripes on the gill covers then are black in the male and blackish red in the female. Based on their slender build and their small head, it is assumed that they are bubblenest builders, but good observations are not available.

Betta imbellis Ladiges, 1975
LENGTH: 2¼ in. (5.5 cm).
FIN RAY AND SCALE COUNTS: 0–I/7–9; A, III/22–25; C 11; L.l. scales 27–30.

DISTRIBUTION: Malay Peninsula. It occurs in swampy waters, canals, and flooded rice paddies. According to Linke, the water temperature at one locality was 93°F (34°C), the water hardness 8 to 10 DH, and the pH value about 7.0. The name of the species *imbellis* (not warlike) unfortunately is often translated as "peaceful fighting fish." It is possible that other *Betta* species (for example, *B. splendens*) are more aggressive but *B. imbellis* is by no means a docile species.

DESCRIPTION: The basic color of the adult male is reddish brown; under light, it is iridescent blue. The caudal fin has a wide red edge. Ventral fins and the posterior tip of the anal fin are red. In fully mature males, the ventral fins have white tips. The females are paler and usually show four vertical stripes. During the spawning period the stripes grow darker and blend in with the body color, disappearing almost entirely.

MAINTENANCE AND BREEDING: Kept in a small container, the males of *B. imbellis* are just as aggressive toward males of their own kind as most other bubblenest-building *Betta* species. Several males can be kept together in a large well-planted aquarium.

Ripe females are recognized by their easily visible whitish genital papilla, in addition to their larger abdomen. In one spawning a total of 80 to 150 eggs is laid. With adequate feeding, one breeding pair can spawn four to five times in eight- to ten-day intervals. According to several current reports, the females are said to participate in carrying the eggs to the nest. The eggs hatch at a temperature of 79°F (26°C) after thirty-five hours, and the young fish are free swimming on the third or fourth day after spawning. The best food for initial feeding is an abundance of *paramecia* or rotifers. The young grow very quickly and are ready to spawn at an age of three to four months.

Betta smaragdina Ladiges, 1972
LENGTH: 2¼ in. (6 cm).
FIN RAY AND SCALE COUNT: D, I–II/7–9; A, IV–V/22–26; L.l. scales, 31–35.
DISTRIBUTION: Eastern Thailand. Schaller claims he caught the fish in the vicinity of Korat (Naklion Ratschsima) in an almost dried-up pond. The fish imported by Schaller in 1970 were spawned by E. Roloff.
DESCRIPTION: The following data are based on a pair I received from Roloff's brood. The adult male is dark olive-brown. There is a lighter spot on each scale. These spots are a light-green iridescent color, particularly on the sides of the body. Gill covers and the soft tissue of the dorsal and pelvic fins likewise have a greenish coloration. The anal fin is wine-red to red-brown. The pelvic fins are red with white tips. The female shows similar coloration, except that all hues are considerably duller. If the fish are threatened or

stressed, four vertical stripes appear. This species does not seem to be particularly productive. Scarcely more than 100 eggs are laid per spawning. The young fish take surprisingly coarse food and can be fed immediately with newly hatched brine shrimp nauplii.

Betta splendens Regan, 1909

COMMON NAME: betta, Siamese fighting fish.

SYNONYMS: *Macropodus pugnax* var. Cantor, 1850; *Betta pugnax* Bause, 1897; *Betta pugnax* var. *rubra* Köhler, 1906.

LENGTH: 2¼ in. (6 cm).

FIN RAY AND SCALE COUNT: D, I–II/8–10; A, II–V/21–26; Pec. 5; Pel. I/5; C 11; L.l. scales, 27–31.

DISTRIBUTION: Southern Vietnam, Kampuchea, Thailand, and the Malay Peninsula. Occurrences in Laos and Burma probably came about by releases. In nature, *B. splendens* can be found in many types of freshwater habitats, including both standing waters and running waters. It is also found in flooded rice paddies and their accompanying irrigation systems.

When aquarists speak of "fighting fish," they usually mean *B. splendens*. This species was not described scientifically until 1909, yet it was involved in the aquarium trade much earlier. It was first imported to France in 1874 under the name *B. pugnax*. At that time nobody was able to breed this fish. It was not until 1892 that Jeunet succeeded in breeding *B. splendens* for the first time in Europe. *B. splendens* was one of the first tropical aquarium fish to reach the shores of the United States on a regular basis.

DESCRIPTION: Under existing political conditions in Southeast Asia, it is impossible to determine what *B. splendens* looked like originally. Different geographic forms in Thailand intermingled with released domesticated fish so often that the original appearance of the species can only be guessed at. According to Ladiges and Vierke, there are, for example, in Thailand considerable differences between the "wild Bettas" found in the vicinity of large metropolitan areas and the fish that occur in more sparsely populated areas of the country. Bettas found in the klongs of Bangkok differ considerably from bettas that come from the waters of the Korat Plateau, which is in the vicinity of the border of Kampuchea, or from the bettas of the lowlands of the Menam River. These differences are seen both in the shape of the body and in the number of anal fin and dorsal fin spines. Since the breeding of domesticated Siamese fighting fish for fighting purposes has a long tradition in Thailand—to which were added the long-finned Siamese fighting fish from about 1910 on—it cannot be ruled out that some of the wild stock in Thailand are actually derived from released fish. These wild forms of *Betta splendens* are rarely found in the aquarium trade. It is interesting to note that the fish called Siamese fighting fish all over the world is called "pla kat khmer" by the Thai, in which "khmer" means the country of Khmer (i.e., Kampuchea).

MAINTENANCE AND BREEDING: Siamese fighting fish or bettas can be kept in almost any size aquarium, where they can be bred rather easily in a variety of different water conditions. In most localities normal tapwater can be used in breeding this species. Individual males or individual females can be kept together with many other kinds of fishes. A pair, however, usually cannot be kept together, with or without the presence of other fishes, because the male will often kill the female, except during breeding. A whole school of females can be kept with other fishes, but only one male should be kept in a community aquarium, because males instinctively fight with one another, often to the death of one. They usually get along perfectly well with most other fishes but not with each other. One must be careful in choosing fishes with which to keep a lone male betta, because those with a tendency to be fin nippers will almost constantly pick on the long flowing fins of a male betta, which could bring it no end of problems.

Siamese fighting fish possess a well-formed labyrinth organ and can survive without harm in surprisingly high temperatures. They should not be kept below 78°F (25°C), and for breeding the temperature should be raised to 83° to 86°F (28° to 30°C). Small aquariums of about five gallon capacity are adequate for breeding. An abundance of vegetation should be provided to give the female plenty of hiding places. Java moss, *Vesicularia dubyana,* is quite suitable for this purpose. A breeding aquarium should not be aerated very strongly or filtered by conventional means, although a very slowly operating sponge filter will do no harm and will help keep the water clean.

The bubblenest of *B. splendens* consists of relatively large air bubbles and is affected by water currents. Whereas the larvae hatched from floating eggs develop normally without a bubblenest on the surface, sinking eggs and newly hatched larvae of the bubblenest-building bettas must be suspended in the bubblenest until the young fish are free swimming. If they are

TOP: Red form of the Siamese fighting fish (*Betta splendens*). Long-finned forms are bred in aquariums; they are not the same as the breeds used in Thailand for fish fighting tournaments. BOTTOM: A pair of *Betta smaragdina*. The male displays in front of the female by spreading his fins.

not held close to the surface, such larvae do not develop normally. The differentiation of the sexes is simple, since the males usually have more intense colors than the females (unless they are special females bred for color intensity) and always have larger fins. The females of long-finned bettas frequently show three darker horizontal stripes. Their genital papilla can almost always be easily recognized as a small rise of yellowish white directly anterior to the anal opening. Full-grown but not too old fish are best for breeding. Siamese fighting fish grow very rapidly and age rapidly; therefore a large brood cannot be expected from fish that are much older than one year. Older fish will breed, but they are just not very fecund. After the spawning, the female should be removed from the breeding aquarium. The young fish swim freely on the fourth or fifth day after spawning. They are about 3 to 3.5 mm long and are best fed initially with *paramecium* or rotifers.

BREEDING FORMS: *Betta splendens* has been bred in most of the Orient for its fighting ability for over 120 years, and in Kampuchea (formerly Cambodia) even longer. These special breeds are used in staged fights upon which wagers are made. Aquarists have had little other than academic interest in these fighting fish and the sport they provoke. But there has been a great deal of interest in the long-finned mutants that originally showed up in fish that were destined for staged fights. These long-finned mutants became known in Thailand around the turn of the century. They arrived in the West in the middle 1920s, a blue-red strain being the first import in 1925. One year later a Hamburg importer was able to offer a pure blue strain of this long-finned Siamese fighting fish. These were soon followed by pure red fish, which were finally followed in 1932 by the first emerald green fish. Today there are long-finned bettas in many hues. In addition to light-red and dark-red fish, there are light-blue and dark-blue fish, emerald green, yellow, yellow-red, slate colored, brown, albinos, and even black bettas.

Mouthbrooding *Betta* species

The mouthbrooding species of the genus *Betta* are found more frequently in running waters than are their bubblenest-building relatives. Some of these mouthbrooders, such as *Betta picta* and *Betta unimaculata*, are even found in mountain streams. It seems likely that mouthbrooding in these fishes may have evolved as an adaptation to living in waters swift enough to make bubblenest building a very difficult task. There is, however, no fossil evidence to support this theory, and contrarily, some mouthbrooding *Betta* species are found in waters that also support bubblenest builders.

Mouthbrooders can often be distinguished from bubblenest builders by the legnth of the head in relation to body length, the heads of mouthbrooders generally being proportionately longer than the heads of bubblenest builders. Also, mouthbrooders are more inconspicuously colored than bubblenest builders. Except during the spawning period, it is difficult to distinguish the sexes of the mouthbrooders, because the differences in color pattern and fin shape are insignificant. The behavior of mouthbrooding species differs in many ways from that of the bubblenest-building species. Mouthbrooding male bettas do not establish firm, fixed territories; their territorial behavior is quite subdued compared to that of bubblenest builders. In fact, these fishes seem to be rather peaceful toward others of their own kind, except for a short time directly before and during spawning activity.

During spawning, which usually occurs near the bottom of the aquarium, the male embraces the female without turning her upside down, as is done by most of the bubblenest-building bettas. The first real spawning is usually preceded by several pseudo-spawnings. After each embrace, it is the female who first releases herself, while the male remains motionless, with his caudal fin held at a right angle to the body. The eggs released during the embrace come to rest against the bent caudal fin of the male. In each embrace anywhere from two to fifteen eggs are laid. The eggs on the male's caudal fin are taken into the mouth of the female. The pair then meets head to head, and the female spits out the eggs. Both fish catch the eggs in their mouths. The eggs caught by the female are

TOP LEFT: *Betta splendens* (male), the wild form of the Siamese fighting fish. The males have vertical stripes on the gill covers which are clearly visible when the fish is excited. The fish shown is from a population in south Thailand and was collected by P. Nagy. It is difficult to know if this is what the original wild *B. splendens* looked like.

TOP RIGHT: *Betta coccina*, male.

CENTER LEFT: *Betta foerschi*, male in courting color (dark with golden vertical stripes on the gill covers).

CENTER RIGHT: *Betta anabatoides*, male.

BOTTOM LEFT: *Betta pugnax*, male, from the island of Penang (Malaysia).

BOTTOM RIGHT: *Betta edithae*, male. It was first assumed that this fish, discovered on Borneo in 1978, was a form of *Betta taeniata*. It has now been established as a new species, recently described as *Betta edithae* Vierke, 1984.

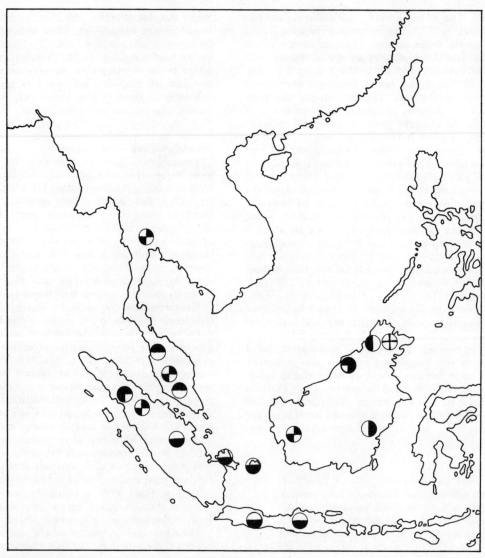

Distribution map for mouthbrooding *Betta* species

◖ *Betta anabatoides* ◓ *Betta pugnax*

◕ *Betta balunga* ◔ *Betta rubra*

◕ *Betta macrostoma* ◓ *Betta taeniata*

◒ *Betta picta* ⊕ *Betta unimaculata*

then spit out again, and this "game" continues until all of the eggs are securely stored in the mouth of the male. As long as the female still has eggs in her mouth, no new embraces are attempted. The whole spawning sequence can last for hours. If the fish are disturbed during the spawning by other fishes in the aquarium, it is the female who chases away the intruders.

The egg-carrying male usually remains close to the surface of the water, at an oblique head-up angle. Under strong back lighting one can see the eggs through the gill covers and the thin tissues of the throat. Typically representative of mouthbrooding *Betta* species, the young of *B. taeniata* emerge from the eggs in about four days at a temperature of 77°F (25°C). After another four or five days their yolk sac is consumed, and they are released from the male's mouth. At this time the larval fish are about ¼ in. (5 to 6 mm) long. They can be fed newly hatched brine shrimp nauplii right from the start. This development proceeds in a similar manner in other mouthbrooding species. Minor species differences can be seen in the rate of development, but most of those differences are at least partially temperature dependent.

Successful spawnings of mouthbrooding bettas in the aquarium are not very numerous, partly because these biologically interesting but somewhat drab-looking fishes are imported only rarely and also because they simply do not reproduce well in captivity. *Betta taeniata* is the most frequently spawned of these fishes. Only occasional successful spawnings of other mouthbrooding species have been reported. Properly kept fish seem quite willing to spawn, but the eggs often disappear after only a few days of oral incubation; they are probably eaten by the brooding parent fish. Presumably these fish are lacking in one or more of the environmental conditions necessary for normal brood care. Since most of these species have been imported and bred so rarely, it is probable that we simply have not yet learned enough about them to be able to breed them at will, as we do with some of the other *Betta* species. As in any other difficult-to-breed fish, a basic prerequisite for successful breeding of these fishes is the proper conditioning of the breeders. These fishes are generally heavy eaters that need nutrient-rich but not fatty foods. An excellent food for bringing these fish into spawning condition is live mosquito larvae. Unfortunately, in temperate climates they are available only seasonally. A good substitute is tubifex worms or live adult brine shrimp. Another prerequisite for successful breeding is placing the fish in the proper environment. Most of the mouthbrooding bettas available to the trade are wild-caught specimens. In the aquarium these are often shy and require lots of hiding

places. Because of their shyness, the best breeding results will be achieved by keeping each pair separate from other individuals. If a large number of these fishes, especially *Betta taeniata*, are kept in one aquarium it would not be unusual for only one male to take on the color changes typical of the male, while the other males show the same coloration as the females. Similar phenomena would be expected in other mouthbrooding bettas as well. Some mouthbrooding species come from the rapidly flowing coolish waters of higher elevations and should not be kept at too high a temperature. Specific suggestions on care are given in the species descriptions.

The taxonomic classification of mouthbrooding *Betta* species is an extremely difficult task. It is suspected that several species have been described twice. The mouthbrooding *Betta* species, according to Vierke, belong to two different basic types. He has temporarily designated these fishes as the "taeniata type" and the "pugnax type" and has described several differences between these two basic types. For example, he said that the maximum length of the taeniata type is about 3 in. (8 cm) and that of the pugnax type 4 in. (10 cm) or more. In the taeniata type he described the anal fin of an adult male as reaching backwards only as far as the anterior portion of the caudal fin, whereas in the pugnax type the tip of the anal fin reaches the posterior part of the caudal fin. The caudal fin of the taeniata type, according to Vierke, is rounded and because of slightly extended fin rays often appears serrated. In contrast, only the central rays of the caudal fin of the pugnax type are extended, thus giving the fin a pointed appearance. Vierke claims that males of the pugnax type have elongated pelvic fins, whereas males of the taeniata type have shorter pelvic fins. There are several assumptions and theories on the classification of the mouthbrooding *Betta* species, which, for lack of space, cannot be dealt with in this text. Based on what is presently known about these fishes, Vierke's assumption that there are only two species of mouthbrooding bettas and that these should be subdivided into numerous subspecies seems to me most logical. However, in the descriptions that follow an older classification system that recognizes several distinct species is used. A comprehensive taxonomic revision of the entire *Betta* complex is not expected in the near future.

Betta anabatoides Bleeker, 1850
LENGTH: Up to 4¾ in. (12 cm).
FIN RAY AND SCALE COUNT: D, I/7–9; A, II/25–30; P, 13, Pel. I/5; L.l. scales, 31–34.
DISTRIBUTION: The fish on which the original descrip-

tion was based came from southeast Borneo. The species is also said to occur on Sumatra, Sumba and in the vicinity of Singapore. In all collection sites outside of Borneo it is doubtful whether the fish discovered there were indeed *B. anabatoides.* The fish brought back by W. Foersch from Borneo in 1978 were probably the first of this species brought back alive. In 1979 Vierke called attention to the fact that the fish examined by Kühme in 1961 for their behavior could not have been *B. anabatoides,* as was stated, but most likely were *B. taeniata,* which, of course, does not lessen the value of Kühme's comparative study between bubblenest-building and mouthbrooding *Betta* species.

DESCRIPTION: *Betta anabatoides* is a relatively stocky looking mouthbrooder that has a large head and a caudal fin that is somewhat pointed. The basic background color is brownish, with an irregular darker striped pattern on the lower side of the body. During courtship and spawning, the color of the male is somewhat brighter.

MAINTENANCE AND BREEDING: Soft water and temperatures between 75° and 80°F (24 to 27°C) are recommended. The breeding of this species has been described as typical of a mouthbrooding member of the genus.

Betta balunga Herre, 1940
LENGTH: Up to 2¾ in. (7 cm).
FIN RAY AND SCALE COUNT: D, I/7; A, 28–29; L.l. scales 30–31.
DISTRIBUTION: Northern Borneo, where this fish was found in the Balung River. The species has not been imported alive to date.
DESCRIPTION: Nothing is known about the coloration or specific breeding habits of living fish.

Betta macrostoma Regan, 1909
LENGTH: Up to 4 in. (10 cm).
FIN RAY AND SCALE COUNT: D, 11; A, 26; L.l. scales, 32. This species has no spines.
DISTRIBUTION: This species was first described by Regan, based on a single preserved specimen, and it was known only from north Borneo. Live specimens of this species were not known. Recently, however, this species was found in the Sultanate of Brunei and brought back to the United States, where it was successfully spawned in an aquarium. The male carries the eggs in his mouth for seven to ten days, until the young fish are free swimming.
DESCRIPTION: A noticeably deep cleft mouth and a distinctive peacock spot on the posterior part of the dorsal fin are typical of this species. Juveniles and subadults have two rows of spots on the body.

MAINTENANCE AND BREEDING: The fish were found in running water, an environment that is on the coolish side and rather high in oxygen. *Betta macrostoma* is an excellent jumper that can move over land for some distance. Keeping the fish therefore necessitates a tightly covered aquarium. Breeding is typical for a mouthbrooding *Betta* species.

Betta picta (Cuvier and Valenciennes, 1846)
SYNONYMS: *Panchax pictum* Cuvier and Valenciennes, 1846. *Betta trifasciata* Bleeker, 1850.
LENGTH: Up to 2¼ in. (5.5 cm).
FIN RAY AND SCALE COUNT: D, I/6–8; A, II/18–22; P. 12; L.l. scales, 28–30.
DISTRIBUTION: Java, Sumatra and its offshore islands of Bangka and Belitung. Its recorded occurrence on Borneo and the Malay Peninsula have not been confirmed. This species was also found on Java at an elevation of about 5000 feet (1500 m). This species was first imported to the West in 1935.
DESCRIPTION: Its background color is light brown, and it has three darker horizontal stripes and several darker spots on the forward part of its back.
MAINTENANCE AND BREEDING: It is best kept in soft water at temperatures of 68°F to 75°F (20 to 25°C). It has been bred several times in the aquarium, in the typical mouthbrooder fashion. Growth is rapid, and adult size is reached at an age of six months.

Betta pugnax Cantor, 1850
SYNONYMS: *Macropodus pugnax* Cantor, 1850. *Betta brederi* Myers, 1935. (*Betta brederi* was probably only a local form of *Betta pugnax* from Johore, in the southern part of the Malay Peninsula. It differs from *B. pugnax* only by a more greenish background color.
LENGTH: Up to 4 in. (10 cm).
FIN RAY AND SCALE COUNT: D, I/7–9; A, II/22–26; P, 12; Pel, I/5; C, 11, L.l. scales, 28–30.
DISTRIBUTION: The species was originally described from specimens captured on the island of Penang, which is located off the western coast of Malaysia. It also occurs in many other places on the Malay Peninsula. It is found in nature in a variety of different habitats and is reported to occur even in rapidly flowing brooks. It was first imported in 1905; subsequent importations have been irregular and in small numbers.
DESCRIPTION: The color is variable and strongly dependent on environmental conditions and activity. If the fish are healthy, their background color is reddish brown and on each scale there is an iridescent greenish or bluish spot. An iridescent green color covers the throat area and the outer margin of the anal fin. Under stressful conditions these fish are yellowish brown and show eight or nine vertical stripes.

MAINTENANCE AND BREEDING: *Betta pugnax* should not be kept in hard water. A temperature of 73 to 76°F (23 to 25°C) is sufficient. This species will not fare well in hard water or at too high a temperature. Spawnings of this mouthbrooder have been reported several times. If the parent fish are well fed with a good variety of foods, they do not bother their young. They grow quickly and mature in six months.

Betta rubra Perugia, 1893
LENGTH: Up to 2 in. (5 cm).
FIN RAY AND SCALE COUNT: D, I/7; A, III/21; L.l. scales, 30.
DISTRIBUTION: It is known from Toba Lake on Sumatra at 3000 feet (900 m) above sea level. It is a species of uncertain taxonomic status, which has not been imported alive to date. It is mentioned here because in older aquarium literature wild *B. splendens,* having a reddish primary coloration and found in the vicinity of Singapore, have often been described erroneously as *"Betta rubra."*

Betta taeniata Regan, 1909
SYNONYMS: *Betta trifasciata* Karoli, 1882; *Betta macrophthalma* Fowler, 1934.
LENGTH: Up to 2¾ in. (7 cm).
FIN RAY AND SCALE COUNT: D, I/7–9; A, II/20–25; P, 12–14; C, 11, Pel, I/5; L.l. scales, 28–30.
DISTRIBUTION: This fish, first described by Regan, came from Borneo. Other occurrences are known from Sumatra. This species also occurs on the Malay Peninsula and in Thailand. Linke found it about 50 mi (80 km) north of Singapore. *Betta taeniata,* as Tweedie states for the occurrences in Malaysia, is in no way identical to *B. pugnax.*
DESCRIPTION: Primary coloration is yellowish brown to brown; certain specimens are also reddish brown. Two darker horizontal stripes are almost always recognizable. One stripe runs from the tip of the mouth through the eye to the root of the tail. The other stripe begins at the edge of the gill cover and ends in the lower part of the tail stem. Sexual differences are difficult to determine; ripe females are recognized by the greater roundness of the belly.
MAINTENANCE AND BREEDING: In nature this species lives in both running and standing water. It is very stable in an aquarium and probably easiest to spawn of all mouthbrooding *Betta* species. It is best kept in a well-planted aquarium at a temperature of about 77°F (25°C). Fish are healthiest in water that is not too hard and is slightly acidic. They are quite peaceful toward their own kind as well as toward other kinds of fishes, so keeping several specimens together presents no special problems. Spawning can be successful if one removes from the aquarium all other fish as soon as the gathering of eggs is observed, so that only the egg-carrying male remains. The ¼-inch-long (5 to 6 mm) young fish which are released from the mouth after eight to twelve days can be fed immediately with newly hatched brine shrimp nauplii.

Betta unimaculata (Popta, 1905)
SYNONYMS: *Parophiocephalus unimaculatus* Popta, 1905; *Betta ocellata* De Beaufort, 1933.
LENGTH: Up to 4½ in. (11 cm).
FIN RAY AND SCALE COUNT: D, O–I/6–11; A, O–I/26–33; L.l. scales, 31–35.
DISTRIBUTION: This species is found in northern Borneo, in a variety of different habitats. According to Inger and Chin, these fish are found in swift streams, where they overcome small rapids and other obstacles by leaping over them. They also appear capable of short overland migrations as a means of getting around larger obstacles. They have been found above waterfalls as well as in small puddles connected only by rapidly running trickles.
DESCRIPTION: The primary coloration is yellowish to brownish with iridescent greenish scales on the sides of the body. Differentiating the sexes in immature fish is not easy. In mature fish, however, the males are a little larger, and during courtship they show iridescent yellowish green color on the gill covers and on their sides. Females show the same colors but are not nearly as brilliant.
MAINTENANCE AND BREEDING: These fish are very agile leapers, and their aquarium should therefore be kept tightly covered. They live peacefully together and with other kinds of fishes. In their rare disputes they usually go no further than offering an aggressive display when faced by an opponent.

They differ somewhat from other *Betta* species in their spawning behavior. As soon as the eggs are released, they are taken up directly by the male, without the aid of the female. The young fish are released from the mouth of the male after nine to ten days at a temperature of 77°F (25°C). The young take newly hatched brine shrimp nauplii immediately and grow very quickly.

Genus *Ctenops*
McClelland established this monotypic genus specifically for this peculiar fish in 1844. Bleeker placed the species *Tichopsis vittatus* (under the name *Osphromenus vittatus*) in the genus *Ctenops* in 1877. This incorrect classification led to *Trichopsis vittatus* being mistakenly called *Ctenops vittatus* in aquarium hobby literature.

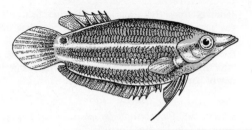

Ctenops nobilis

Based on the few specimens imported alive to date information about the keeping of this fish is incomplete.

Genus *Malpulutta*

Malpulutta is a monotypic genus established by Deraniyagala in 1937, the distribution of the single species—*M. kretseri*—is limited to Sri Lanka.

Ctenops nobilis McClelland, 1844
SYNONYM: *Osphromenus nobilis* Day, 1876.
LENGTH: Up to 4 in. (10 cm).
FIN RAY AND SCALE COUNT: D, V–VII/7–8; A, IV–V/24–28; L.l. scales, 29–33.
DISTRIBUTION: Occurrences are known from the lowlands of India along the lower reaches of the Brahmaputra River and from the delta areas of the Brahmaputra and the Ganges. The fish seems to be rare throughout its range. The fish was first imported in 1912. They have been imported only rarely since then.
DESCRIPTION: Characteristic of this peculiar labyrinthfish are its dorsoventrally compressed head, its pointed snout, and its oddly located dorsal fin which is attached very far to the rear on the back. Its primary coloration is brown. An interrupted silvery horizontal stripe stretches from the eye to the center of the caudal peduncle. A similar stripe runs from the pectoral fin, along the side, and a third stripe extends along the base of the dorsal fin. In some individuals there is a black eyespot surrounded by a halo of lighter color on the upper part of the caudal fin. According to Ladiges, the young fish go through a multiple color metamorphosis during their growth. Young fish ½ in. (10 to 15 mm) long have a wide, beige vertical stripe posterior to the head bordered in front and back by a dark brown zone. The anal and caudal fins are also beige. At 1 in. (15 to 25 mm) young fish are uniformly dark, with about 15 lighter but indistinct horizontal stripes. Larger young fish show a washed out speckled pattern against which two small oblique stripes clearly emerge. One stripe starts at the anterior base of the dorsal fin and ends at the base of the pelvic fin; the other stripe runs from the back base of the dorsal fin to the base of the tail fin.
MAINTENANCE AND BREEDING: Nothing is known about the biology of this species, particularly its reproduction. Considering its taxonomic position in relation to the *Trichopsis* species, and its geographic distribution, it can only be assumed that it is a bubblenest builder. This species is said to be very frail in an aquarium.

Malpulutta kretseri Deraniyagala, 1937
COMMON NAME: Spotted Pointed Tail Gourami
LENGTH: Males up to 3½ in. (9 cm) of which over one-third is the caudal fin. Not all males have such long tail fins. The females do not grow any longer than 2 in. (5 cm).
FIN RAY AND SCALE COUNT: D, VIII–X/4–6; A, XIII–XVII/7–11; P, 12; Pel, I/5; C, 13–15; L.l. scales, 29–30.
DISTRIBUTION: This species is found both in the northwestern and the western parts of Sri Lanka. Rolf Geisler and Alfred Radda found them in small, heavily overgrown habitats in the Kottawa Forest Reservation on the slopes of the southern highlands in the southern part of the island. According to Geisler, they were found together with *Belontia signata* and various *Puntius* and *Rasbora* species in irrigation ditches and brooks. In nature this species occurs in mineral-poor, slightly acidic water. The measured values in different places where they have been found lie between 0.2 and 4 DH (total hardness) and between 6.2. and 6.6. pH. Water temperatures ranged from 77 to 82°F (25 to 28°C).
DESCRIPTION: Deraniyagala divided the species *Malpulutta kretseri* into two subspecies in 1958. One is more reddish and a little larger than the other; it is considered to be the nominate form. A smaller form from the province of Sabaragamuva, has a bluish primary coloration. It received the name *M. kretseri minor*. As far as is known, only fishes of the red form have reached the West alive. Whether the subdivision mentioned is justified cannot, for lack of space, be discussed here. Similar phenomena occur in other species; that is, the same species occurs in different parts of Sri Lanka in various color variations, in each case a more reddish form and a more bluish one. Similar color forms are known, for example, for *Belontia signata*, *Puntius titteya*, and *Rasbora vateri floris*.
MAINTENANCE AND BREEDING: *Malpulutta kretseri* are easily kept in an aquarium when there are no other species in the tank, except perhaps other small laby-

rinth fishes such as the *Trichopsis* species. *Malpulutta kretseri* is a very cautious and shy fish that spends much of its time in sheltered areas of the aquarium. If these fish are kept in a well-planted aquarium, one does not get to see much of them. Imported fish first take only live food and become used to nonliving food only after quite some time.

This species sometimes spawns in caves or other sheltered areas of the aquarium. A bubblenest is usually built, either in a cave or closer to the surface under overhanging plant leaves. If there are floating plants on the surface, the nest is built under these most of the time. A large number of young should not be expected, but the young fish do grow quickly and are easy to raise. Courtship can take several days. Spawning is initiated by the female who indicates her readiness to spawn by nudging the male on his side with her mouth. During the embrace the female as a rule ends up in a sidewise position, and thus embraced, the pair sinks to the bottom. The released and fertilized eggs (sinking eggs) at first remain on the anal fin of the male. Only when the fish release themselves from the mating do the eggs slide along the anal fin to the bottom of the aquarium. There they are picked up in the mouths of both partners and carried to the nest. The female is usually the more active egg collector, and she may begin collecting the eggs while they are still lying on the anal fin of the male. During each embrace four to eight eggs are released, and pseudo-embraces often occur between productive embraces. The larvae emerge from the eggs after about 48 hours at a temperature of 77°F (25°C), and six days after spawning the young fish are free swimming. The parents do not bother their offspring very much, so if one is not after particularly large numbers of offspring, the young fish can be kept with their parents. This is only true if the aquarium is not too small and also is well planted. If numerous young are desired, either the parents must be removed from the breeding aquarium, or the young fish swimming near the surface must be carefully skimmed off and transferred to a special rearing aquarium. Brine shrimp nauplii can be fed to the young right away.

This species depends very little on its labyrinth organ for oxygen, and these fish can remain in a protected spot in water with normal oxygen saturation for hours without breathing atmospheric oxygen. If the fish are threatened, they often try to escape by jumping. They jump out of the aquarium through the most narrow slits or openings, and therefore the aquarium must be very tightly covered.

The spotted pointed tail gourami is a fish that thrives only in an aquarium especially equipped for it. It is an extremely interesting fish for the true specialist.

Genus *Parosphromenus*

The genus *Parosphromenus* established by Bleeker in 1879, contains four species, of which two are new species from Borneo, which were recently added to the two earlier-known species. In 1859, based on one specimen, Bleeker described a tiny labyrinth fish from the island of Bangka as *Osphromenus dreissneri*. This dwarf fish seemed so peculiar to Bleeker, and so different from all other labyrinth fishes known at that time, that in 1879 he established a new genus for it, *Parosphromenus*. For decades the Bleeker specimen remained the only known specimen of the species. This fish was rediscovered by Herre 75 years later. It has since been established that this species is not limited only to the island of Bangka but also occurs on Sumatra and on the Malay Peninsula. More than 90 years after the discovery of the first species of *Parosphromenus* Tweedie found the second species of the genus, *Parosphromenus paludicola,* on the Malay Peninsula. Since then, two more species of *Parosphromenus* have been discovered on Borneo and brought back alive to Europe by Foersch, Korthaus, and Hanrieder. The history of the discovery of these four species of tiny labyrinth fishes suggests that it can be very difficult to discover such small and almost sedentary fishes in open waters. It is quite possible that additional species of this genus will be found.

In water having a normal percentage of oxygen saturation these dwarf fish easily get along without using their labyrinth organ to extract atmospheric oxygen. Foersch was able to show, for example, that *P. dreissneri* survived over 100 days, when its approach to the surface was barred. The fish indicated no discomfort, and the experimental pair spawned ten times during this period. This data suggests that these fish, as with *Malpulutta kretseri,* are dependent only to a limited extent, if at all, on oxygen absorbed through the labyrinth organ.

In nature these species occur predominantly in clear, running water. Their nest-building technique is an adaptation to their biotope. Since running water does not permit the construction of a normal bubblenest at the surface, the *Parosphromenus* species build their bubblenests, as a rule, in caves and other protected locations under the water surface. The development of eggs and larvae is slower in the *Parosphromenus* species than in almost all other labyrinth fishes; at 77°F (25°C) it takes three days for the eggs to hatch. Upon emerging from the eggs the larvae still have a large yolk sac, and it takes another three to four days until they are free swimming. At that point they are about ¼ in. (5 mm) long and are capable of eating newly hatched brine shrimp nauplii, but the fry can also be fed paramecium or extremely fine dried natural food.

Descriptions of Species

Parosphromenus species are relatively long-lived fishes, which remain fertile even in their third year of life.

Parosphromenus dreissneri Bleeker, 1859
COMMON NAME: Splendid dwarf gourami
SYNONYMS: *Osphromenus dreissneri* Bleeker, 1859; *Polyacanthus dreissneri* Günther, 1861.
LENGTH: Up to 1½ in. (3.5 cm).
FIN RAY AND SCALE COUNT: D, XII/7; A, XIII/8–9; P, 2; Pel, I/5; C, 12; L.l. scales, 30.
DISTRIBUTION: This species is known from the Malay Peninsula and the islands of Sumatra and Bangka. Schaller found the species about 50 mi (80 km) northwest of Singapore near the bank of a deep ditch. The fish were near the bottom, hovering in a head-down position, between reed grasses. In 1955 Schmidt found this species on the island of Sumatra, in the vicinity of Palembang, in a swift forest stream, which was up to 6 ft (1.80 m) deep; the fish were found close to the bank. Findings are also known from shallow overgrown waters. Tweedie caught them in such a ditch in the vicinity of Singapore. The species was first imported in 1950. Later imports were made by Schmidt in 1955 and Schaller in 1970. Foersch was able to reproduce this species successfully and distribute the fish in the hobby.
DESCRIPTION: The primary background color is a yellowish olive to ocher, with two wide brown horizontal stripes. During the mating period, these horizontal stripes in the male become black and the space between them becomes a brighter yellow. The upper stripe runs along the upper part of the back, starting on the head above the eye continuing to the upper part of the caudal peduncle. The lower stripe runs from the tip of the snout, through the eye, and along the center of the body to the bottom part of the caudal peduncle. The belly becomes darker and the back lighter. The dorsal, anal, and caudal fins have a dark stripe in their center bordered by a bright blue edge on the top and the bottom. There is also a reddish brown area in the caudal fin between its dark base and the deep black stripe. The pelvic fins have a blue iridescent color and a reddish leading edge. The female is much more drab and only hints at the color pattern described for the male.
MAINTENANCE AND BREEDING: This species is suitable mostly for a single species aquarium established especially for them. The fish thrive well in soft water, but they are not extremely demanding in this respect. Total hardness should not exceed 8 to 10 DH. Small inverted clay flower pots with sides broken out can be used as spawning caves. According to Foersch the adhesion capacity of the eggs depends largely on the hardness of the water; therefore very soft water should be used for breeding attempts, if possible, with a total hardness not exceeding 2 DH. Most of the time *P. dreissneri* deposits its eggs under the cave roof without first building a bubblenest. Following courtship displays by the male the fish embrace under the cave roof. This is initiated by the female, who swims up to the male and touches the side of his body with her mouth, whereupon the male embraces the female by encircling her. The female is not turned over in this embrace. Numerous pseudo spawnings precede the deposition of eggs. Initially, after the pseudo spawnings, only a single egg per embrace is released, and it remains on the anal fin of the male. The male remains in his position until the female picks up the egg in her mouth. Then the female attempts to attach the egg to the roof of the cave. The number of released eggs increases during the embraces that follow, and later three to four eggs are released during each embrace. If the eggs fall from the roof of the cave, both fish reattach them. The eggs that are stuck to the roof of the cave are supported by air bubbles during the further course of brood care. Air bubbles are usually seen on the cave roof two to three days after the spawning. Eggs remaining on the cave roof without later bubblenest support can develop, but such development is the exception; most of the time such eggs disappear (they are probably eaten by the parent fish) after a few days. It is possible to collect the eggs from the cave roof through a thin pipette tube and hatch them in shallow glass bowls (i.e., petrie dishes).

Parosphromenus filamentosus Vierke, 1981
COMMON NAME: Splendid thread fin dwarf gourami
LENGTH: Up to 1½ in. (4 cm), of which 4 to 5 mm is the elongated central caudal fin ray.
FIN RAY AND SCALE COUNT: D, XII–XIII/6–7; A, XI–XII/10; P, 12–13; L.l. scales, 29–30.
DISTRIBUTION: This species has been found only in the immediate vicinity of Banjarmasin in south Borneo.

TOP LEFT: Bubblenest of *Betta imbellis*.
TOP RIGHT: Head of a *Betta pugnax* type from Malaysia.
CENTER LEFT: *Betta pugnax* (male), from southern Thailand.
CENTER RIGHT: *Betta edithae* from Borneo. The original description was based on the pictured specimen together with the male shown on the last plate.
BOTTOM LEFT: *Betta pugnax* type from Malaysia, collected by P. Nagy in 1981.
BOTTOM RIGHT: A central Sumatra *Betta pugnax* that grows to large size.

DESCRIPTION: This fish is very similar to *P. dreissneri*, except that the central rays of the caudal fin are extended well beyond the fin itself, forming a thread-like appendage on the tail. During mating the dark horizontal stripes of the male change to blackish brown color and the spaces between the stripes to a light cream color or almost white. The back is yellowish brown. The dorsal, caudal, and anal fins are brownish at their bases and become lighter toward the margins. There is a thin darker band near the outer margin of these fins and a bluish outermost edge. Toward the center the caudal fin has reddish brown color. Between spawning periods males and females are colored almost alike. Their primary background color then is light yellowish brown with washed out horizontal stripes.

MAINTENANCE AND BREEDING: According to Foersch, the spawning procedure is a little less complicated in this species than it is in *P. dreissneri*. The males of *P. filamentosus* build a regular bubblenest under the cave roof *before* spawning. Bubblenests built under overhanging plant leaves have also been observed. Spawning is preceded by a lengthy courtship in which the fish circle each other in an oblique head-down position. The male shows a brilliant display of colors and spreads out all his fins. The female shows an almost uniform yellowish gray color. The fish are very active during the courtship, continuously changing their positions until finally, often after hours, spawning occurs. Spawning does not differ from that described for *P. dreissneri*. During spawning, if the fish touch the eggs earlier released and perched on the male's anal fin, and knock them to the bottom, the eggs are retrieved and brought back to the nest by both parents. The parents seem only slightly aggressive toward their

young. Some young grow up even if they stay in the breeding aquarium with their parents. For a more productive yield of fry, Vierke recommends that the eggs be pipetted out of the bubblenest and that the young fish be raised artificially, away from the parent fish. In each complete spawning 30 to 100 eggs are produced and the embraces follow in quick succession.

Parosphromenus paludicola Tweedie, 1952
LENGTH: Up to 1½ in. (4 cm). Tweedie, in his original description, gives 29.5 mm as the maximum length. It has since been established that this species can grow larger than that.
FIN RAY AND SCALE COUNT: D, XVII–XVIII/6–7; A, IV/7; L.l. scales, 28–30.
DISTRIBUTION: Collections are known only from the east coast of the Malay Peninsula, in Malaysia. The species was first imported in 1977, and the importer successfully spawned this species in an aquarium.
DESCRIPTION: The primary color is yellowish brown with two darker horizontal stripes which are usually visible only during the spawning period. In this species the dorsal fin is longer and its posterior end is farther back than it is in other *Parosphromenus* species. The caudal fin is diamond shaped and pointed at the center. The dorsal fin is crossed by a light horizontal stripe and is bordered by a bluish white edge. The anal and caudal fins are almost colorless, but the anal fin has a narrow white edge. The pelvic fins are greatly elongated and have white tips. The colors are subject to strong changes, according to environmental conditions and activity.
MAINTENANCE AND BREEDING: As with *P. filamentosus*, this species builds its bubblenests in caves prior to spawning. Development of the eggs, larvae, and young fish is the same as it is in *P. filamentosus*. This fish is, however, much more sensitive to environmental conditions. Breeding attempts should be carried out only in soft water.

Parosphromenus parvulus Vierke, 1979
LENGTH: Under 1¼ in. (3 cm). The largest known specimens were 27 mm; thus this species is the smallest known labyrinth fish.
FIN RAY AND SCALE COUNT: D, X–XI/7; A, VIII–IX/10–11, L.l. scales, 27–29.
DISTRIBUTION: Collections are known from streams in the region of the Mentaya River in southern Borneo. The fish were originally found in a slowly running forest stream in which the water was clear but brownish colored. The temperature in this stream was 75°F (24°C). The pH value of the water was below 4.8.

TOP LEFT: Splendid dwarf gourami (*Parosphromenus dreissneri*), male.
TOP RIGHT: *Parosphromenus parvulus*, male. This is the smallest of all known labyrinth fishes and was discovered only recently.
CENTER LEFT: *Parosphromenus paludicola*, male. On the roof of the cave several clearly pigmented young fish can be seen.
CENTER RIGHT: Splendid threadfin gourami (*Parosphromenus filamentosus*) under a bubblenest, containing eggs, on the roof of a cave.
BOTTOM LEFT: The spotted pointed tail gourami (*Malpulutta kretseri*) is adept at concealing itself against the underside of leaves and thus blending in with its surroundings.
BOTTOM RIGHT: Pointed tail macropodus (*Pseudosphromenus cupanus*).

Descriptions of Species

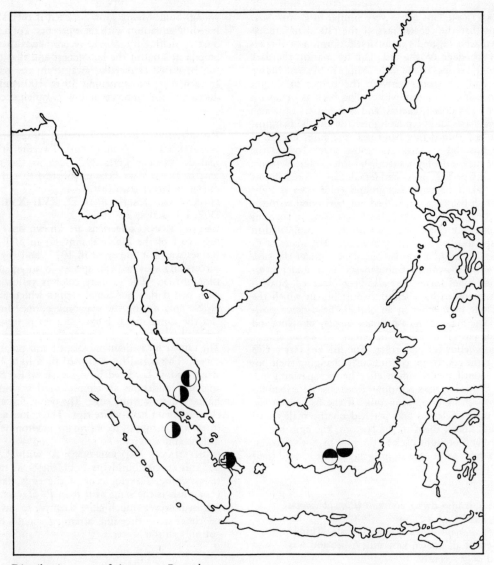

Distribution map of the genus *Parosphromenus*

◐ *Parosphromenus dreissneri*

⊖ *Parosphromenus filamentosus*

◑ *Parosphromenus paludicola*

⊖ *Parosphromenus parvulus*

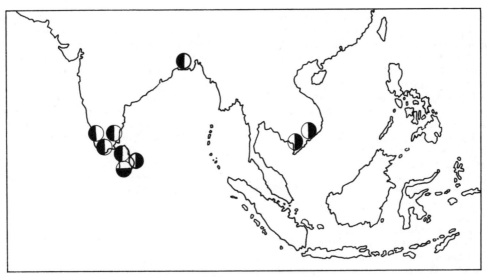

Distribution map of the species of *Pseudosphromenus* and *Malpulutta*

◐ *Pseudosphromenus cupanus*

◑ *Pseudosphromenus davi*

⬬ *Malpulutta kretseri*

DESCRIPTION: The primary background color is brownish red. The dorsal and anal fins are weakly pigmented and have a whitish blue edge. The posterior part of the dorsal fin is dark in the male. The tail fin is colorless. In the female, all fins are colorless.

MAINTENANCE AND BREEDING: Nothing is known about reproduction in the aquarium.

Genus *Pseudosphromenus*

In 1879, Bleeker established a separate genus for the species described by Cuvier and Valenciennes in 1831 as *Polyacanthus cupanus*. He named this genus *Pseudosphromenus* and thus clearly separated the species *cupanus* from the other species of *Macropodus* (Polycanthus). The Dutch physician and naturalist Pieter Bleeker devoted his life to the study of the fish fauna of Southeast Asia and summarized a comprehensive overview of the *Pseudosphromens* species as part of a nine-volume mammoth work, which was published in Amsterdam in 1862 to 1877. As we know today, Bleeker did have valid reasons for separating this species from the genus *Macropodus*. For obscure reasons Regan, in 1909, again put the species *cupanus* back in the genus *Macropodus*. At first described by Köhler as a subspecies, *cupanus* var. *dayi* was placed in the genus *Macropodus*.

Breeders in the aquarium hobby did notice that these small macropodus laid sinking eggs and did not behave like macropodus in any other way. In spite of appropriate references in this regard, *Pseudosphromenus* species were listed until recently in the genus *Macropodus*. Research by Liem in 1963 on the skeletal structure of these fishes and the 1975 comprehensive behavioral studies by Vierke did show, however, that a separate genus for these fishes, which Bleeker had established, was completely justified.

In general, these are lowland fishes that are found in both standing and slowly running waters. *Pseudosphromenus cupanus* is even occasionally found in brackish water. *Pseudosphromenus* species are very versatile in their environmental requirements. Thus H. Randow described an occurrence of *P. dayi* in the highlands of Sri Lanka, where he found this species along the banks of a stream near Bandarawella, at an elevation of 6000 ft (1800 m) above sea level. About this finding Randow wrote, "The water temperature of the creek of Bandarawella was 16°C [60°F], and thorough investigations showed that the pointed tail macropodus did not build bubblenests in this mountain creek but spawned like cichlids on the bottom, on rocks, and on aquatic plants [*Ambulea* species]." In the same study by Randow, locations of collections in the lowlands are described as counterparts to those of the highlands, but the measured temperatures of the lowlands were up to 92°F (34°C). These examples may suffice to show how versatile these fish are.

These fishes frequently establish their bubblenests in caves below the water surface, particularly in slowly running water. Often the nest is established under large overhanging leaves of aquatic plants, especially if there are no floating plants on the surface. During spawning the eggs sinking are gathered up by both the female and the male and carried to the bubblenest. It is not uncommon that the female continues to participate in the brood care. In an aquarium the female normally continues to take care of the brood if the male has been removed. The eggs hatch in about 30 hours at a temperature of 77 to 82°F (25 to 28°C); at 71 to 73°F (22 to 23°C) they need almost 60 hours. The larval fish are free swimming five to six days after the spawning. Paramecium should be given as the initial food. The young fish have rounded caudal fins until they reach a length of nearly an inch (2 to 2.5 cm); then diamond-shaped caudal fins typical for these species develops. The keeping of species of *Pseudosphromenus* is not difficult. They can be bred even in small aquariums and place no demands on the chemical condition of the water. They eat artificial food almost as readily as live food and can withstand temperatures as low as 60°F (16°C) without harm.

Pseudosphromenus cupanus (Cuvier and Valenciennes, 1831)
COMMON NAME: pointed tail macropodus
SYNONYMS: *Polycanthus cupanus* Cuvier and Valenciennes, 1831; *Pseudosphromenus cupanus* Bleeker, 1879; *Macropodus cupanus* Regan, 1909.
LENGTH: Up to 2½ in. (6.5 cm).
FIN RAY COUNT: D, XIV–XVII/5–7; A, XVI–XIX/9–11.
DISTRIBUTION: The lowlands along the Malabar and Komorandel coasts of southern India seem to be the original distribution of this species. It also is found on Sri Lanka. According to Arnold it is also said to be common in Bengal, and according to Hora it also occurs on Sumatra. In 1929 Chevey stated that it also occurs in Tonkin, which is in North Vietnam. Correspondence from Schaller suggests that this statement as well as statements about occurrences in Burma are doubtful. These collectors probably confused these fish with other species. The species abounds in some places and is found in a variety of habitat types, even in brackish water.
DESCRIPTION: The basic background color varies from yellowish brown to brown, with a greenish or reddish shimmer. The color of the male intensifies during the

spawning period, and the female can turn almost black. In the male the dorsal fin is yellowish, and the caudal fin is yellowish and brown or reddish in its lower part. The pelvic and anal fins are reddish, and in some fish the pelvic fins have white tips. There is a clearly defined dark spot on the caudal peduncle, which disappears with age. In males the eye is red. In the nonspawning period it is difficult to distinguish the sexes. But the females are a bit more rotund. This is best seen when one looks at the fish from the top.
MAINTENANCE AND BREEDING: Pointed tail macropodus are peaceful toward other species and among themselves. During brood care the male sometimes builds a second bubblenest and transfers the eggs to it.

Pseudosphromenus dayi (Köhler, 1909)
SYNONYMS: *Polyacanthus cupanus* var. Day, 1878; *Polyacanthus cupanus* var. *dayi* Köhler, 1909.
LENGTH: Up to 2¾ in. (7 cm).
FIN RAY COUNT: D, XIII–XVII/5–7; A, XVI–XXI/10–12.
DISTRIBUTION: As with *P. cupanus,* the distribution of *P. dayi* is not thoroughly known. Certain occurrences are known on Sri Lanka and in South Vietnam. According to other statements, it is also said to occur in southern India and on Sumatra. This last statement is just as doubtful as statements about occurrences in Burma.
DESCRIPTION: This species differs from the *P. cupanus* by a somewhat more slender build and by elongated fin rays in the center of the caudal fin. Both sexes show two dark horizontal stripes on a yellowish brown background. The lower stripe runs from the eye, in a slight downward arch, to the lower part of the caudal peduncle. The other horizontal stripe runs above this one, in about the middle of the upper half of the back, and it ends between the posterior dorsal fin base and the caudal peduncle.

At spawning time the males turn reddish brown on the back and flanks and reddish on the chest and belly. The unpaired fins also show a reddish hue with bluish or greenish iridescent edges. The females turn darker brown at spawning time but not as dark as the females of *P. cupanus.* The sexes are also distinctly different in form. In the males both the dorsal and anal fins are extended to a point. The caudal fin has longer central rays, giving it a central point, and other rays of a blackish blue color that extend beyond the posterior edge of the fin. The pectoral fins are colorless. The elongated rays of the pelvic fins usually have white tips. In the female the dorsal fin is rounded, and there are no elongated rays in the center of the caudal fin.
MAINTENANCE AND BREEDING: *Pseudosphromenus dayi* does not seem to be quite as closely related to *P*

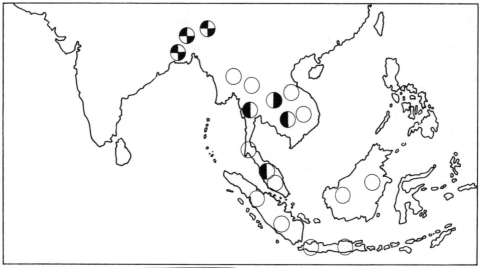

Distribution map of the species of *Ctenops* and *Trichopsis*

◕ *Ctenops nobilis*

◑ *Trichopsis pumilus*

◐ *Trichopsis schalleri*

○ *Trichopsis vittatus*

cupanus as was earlier assumed. Both species can be crossbred without great difficulty. Schaller claims that F_1 hybrids bred with each other produce fertile offspring.

Genus *Trichopsis*

The genus *Trichopsis,* established by Canestrini in 1860, contains three recognized species. These fishes provide an excellent example for showing how difficult it can be to define the concept of a species. *Trichopsis vittatus* inhabits a large part of Southeast Asia. There are considerable differences in markings and coloration within its distribution. An exact description of the different forms (subspecies) has not yet been undertaken. It is obvious, however, that clearly differing forms have been described as species (for example *T. harrisi* Fowler 1934). Some authors also look upon *T. schalleri* as a local form from the area of Korat in Thailand. It has been shown that the distribution of this form is larger than previously assumed, and that it occurs in places together with *T. vittatus*.

The *Trichopsis* species complex does not get any simpler if one tries to bring clarity into the relationships by systematic crossbreeding and backcrossing experiments. Hybrids between *T. vittatus* and *T. schalleri* produce viable offspring when crossed with each other, and according to H. J. Richter, this also holds true for hybrids between *T. schalleri* and *T. pumilus*. With these fish we are faced with the same situation we see in *Macropodus* species (*M. opercularis* and *M. concolor*), or *Colisa* species (*C. fasciata* and *C. labiosa*). It was shown with these fish that they could be crossbred despite easily identifiable differences in form, size, and coloration, and that the offspring of these crossings were capable of reproducing with each other. In order to decode this complicated picture, crossbreeding experiments alone, no matter how valuable they may be, are not sufficient. Beyond that, specimens from the most varied parts of the entire range would have to be available. Nature, throughout the distribution of *T. vittatus,* which includes large parts of Southeast Asia, and runs from Vietnam and Cambodia in the north to Java in the south, has strongly favored the evolution of local forms. Not only does the distribution include numerous islands, but also in the various river systems on the Asiatic mainland the separation of species groups is nearly complete. Such a separation, stretching over a long evolutionary time period, can be the first step on the way toward the formation of a species. Intermediate forms are, therefore, to be expected. Forms whose differentiation is only defined by behavioral or ethological barriers, and in which genetic barriers broken down by unnatural conditions—for example, keeping the forms together in an aquarium.

It is purely an academic question whether groups of animals, which are not separated genetically, but which clearly differ from one another externally, and which do not mate in nature, may be called species. (According to Herre-Röhrs, 1973, a species is a natural, reproducing community with *free choice of mates*. All individuals of a species receive their genetic code from the total genetic stock of the community without themselves having to possess *all the genetic codes* of this community.)

Checking for possibly genetic barriers requires crossbreeding and backcrossing experiments that demand time and space. Proof of behavioral barriers can only be given by means of prolonged observations. Therefore it is obvious that many questions in this regard must remain open for the present. Let it merely be stated here that the number of species of labyrinth fishes would decrease considerably, if all *Macropodus, Colis,* and *Trichopsis* species that produce viable offspring when crossbred were brought together taxonomically. Aquarists should be careful not to equate the conditions in our aquariums with those in nature.

Because of its numerous islands and its river systems, which are often separated from each other by high mountain ranges, the Indo-Malayan area in particular has strongly influenced the splitting of species into numerous local forms (subspecies etc.). Such a breakup into local forms is also more frequent among mammals and birds of this area than elsewhere.

Trichopsis species are found in a great variety of habitats. For example, they are found in extremely turbid water in rice fields and in small irrigation ditches, but they are also found in deeper water.

Adult fish can produce clearly audible sounds. Therefore the genus *Trichopsis* is also called croaking gouramis.

These fishes prefer small living animals to any other food. These are moderately hardy fishes that have been incorrectly described as delicate. They should, however, not be kept together with considerably larger fishes.

Sex determination can sometimes be difficult. If mature females are viewed with strong light behind them, their ovaries can be recognized as a yellowish formation under the air bladder.

All *Trichopsis* species build firm bubblenests, under the cover of floating plants when they are available. Bubblenest building also occurs under overhanging leaves or in hollow spaces between rocks or plants. This is especially true of *T. pumilus.* A particular characteristic of these fishes are the so-called egg packages they release during mating. The eggs from each spawning hang together in one or sometimes two clusters. Thus the male is spared the work of collecting the eggs. Instead he can carry the clustered packet to the nest after each spawning. Occasionally an egg cluster falls apart, and then the female also participates in gathering up the scattered eggs. Such an egg cluster contains four to six eggs and sometimes more. The eggs have a nontransparent white to whitish yellow coloration. The larvae hatch out after 24 to 30 hours at a temperature of 77 to 82°F (25 to 28°C). For another three days the young hang head-up under the bubblenest. After swimming free, the young fish of *T. vittatus* immediately eat brine shrimp nauplii. The young fish of *T. pumilus,* however, must first be fed paramecium or other microorganisms. *Trichopsis vittatus* can be very productive, sometimes producing as many as 300 to 400 eggs in one spawning. However, most of the time the number of eggs laid in one spawning is considerably smaller.

Trichopsis pumilus Arnold, 1936
COMMON NAME: Croaking dwarf gourami
SYNONYMS: *Ctenops pumilus* Arnold, 1936; *Trichopsis pumilus* Stoye, 1984.
LENGTH: Up to 1½ in. (3.5 cm).
FIN RAY AND SCALE COUNT: D, III/7–8; A, V/20–25; L.l. scales, 27–28.
DISTRIBUTION: This species definitely occurs in Kampuchea, Thailand, and on the Malay Peninsula. It is supposedly also found on Sumatra. In 1914 Arnold called this fish *Ctenops* sp. and did not describe it as a valid species until 1936.
DESCRIPTION: Males have larger dorsal and anal fins than females. In the female, the ovaries are clearly visible against a strong backlight. The primary background color is olive-green, darker on the back and lighter on the abdomen. A band of brown to blue color, separated into several spots, runs horizontally across the center of the body, from the tip of the snout, through the eye, to the caudal peduncle. Above and below this horizontal band are rows of tiny bluish iridescent spots. Clearly defined darker spots form a second band on the upper half of the back. The most beautiful feature of these fish are the dorsal, caudal, and anal fins, which are greenish to bluish and on which numerous tiny red spots form a mosaic pattern.

The anterior edge of the dorsal fin and the lower edge of the anal fin have a wine-red border. The posterior part of the anal fin is adorned by a wider area of the same color. The primary colors can differ from individual to individual in this species and can seem either more greenish or more bluish. The pectoral fins are colorless. The elongated pelvic fins are whitish-blue. The eye is blue-green.

BREEDING: The male builds the bubblenest mostly below the water surface. Occasionally mating occurs away from the bubblenest, and the male carries the released egg clusters to the nest.

Trichopsis schalleri Ladiges, 1962
COMMON NAME: Schaller's croaking gourami
LENGTH: Up to 2¼ in. (6 cm).
FIN RAY COUNT: D, III/6–7; A, VIII–IX/19–22.
DISTRIBUTION: The specimens on which the original description was based came from Nam-Mun, in the Korat area, northeast of Bangkok, Thailand. This species apparently also occurs in other areas of Thailand.
DESCRIPTION: The background color is reddish brown to light greenish brown. Two dark horizontal stripes, one along the upper half of the body and one of about the middle of the body, are almost always visible. The dorsal, caudal, and anal fins are bluish with rows of iridescent bluish spots. The dorsal and anal fins have a reddish or reddish brown edge. The elongated pelvic fins are reddish.
BREEDING: Like *T. pumilus,* this species also builds its bubblenest below the water surface.

Trichopsis vittatus (Cuvier and Valenciennes, 1831)
COMMON NAME: Croaking gourami
SYNONYMS: *Osphromenus vittatus* Cuvier and Valenciennes, 1831; *Trichopus striatus* Bleeker, 1859; *Trichopsis striatus* Canestrini, 1860; *Ctenops striatus* Bleeker, 1878.
LENGTH: Up to 2¾ in. (7 cm).
FIN RAY AND SCALE COUNT: D, II–IV/6–8; A, VI–IX/19–28; L.l. scales, 28–29.
DISTRIBUTION: This species is widespread from Vietnam, Kampuchea, and Thailand in the north to Java in the south; it also occurs on Borneo. Within this large area the color patterns of this fish are quite variable. Significant differences also occur in the background color.
DESCRIPTION: The body is laterally compressed, and the head is dorsoventrally flattened. The dorsal fin is short, pointed, and attached far to the rear. The caudal fin is rounded and pointed at the center. The anal fin is large, especially in older males. The background color is reddish brown. Four dark horizontal stripes

run along the body; their form differs according to the fish's locality. Some fish have very regular and clearly defined stripes; in others the stripes break up into more or less associated spot patterns. Fish from Thailand often have only three horizontal stripes and a black shoulder spot. It is difficult to categorize the different color patterns geographically, and authors who have tried to do so have arrived at rather different conclusions. There is a basis for the idea that different color patterns, at least in some geographical areas, occur close to each other. In reflected light the fish shows iridescent hues of greenish and reddish color. The unpaired fins can show both bluish violet or reddish iridescent hues against numerous noniridescent red spots. The edges of the fins can be blue to whitish-blue or even red. The elongated pelvic fins are yellowish or white, and the pectoral fins are colorless. The eye is light blue. The dorsal and anal fins seem to keep growing with advancing age in males, and old males can have very large fins. In adult fish the sexes are easily distinguished by the larger fins of the males and by their more intense colors. In subadult fish the sexes can be a little more difficult to distinguish.

Family Helostomatidae

Genus *Helostoma*

Helostoma is an evolutionarily old genus consisting of only one species. It was established by Cuvier and Valenciennes in 1831. This fish has gill rakers that are effectively used as a sieve for catching very fine plankton. The mouth is equipped with thick protuberant lips that can be pushed forward, and the thick lips bear several rows of movable teeth. Thus the lips are extremely suitable for grazing on algae which, in addition to plankton, represent an important part of its food. Aside from these "lip-teeth," there are no other teeth; thus the fish differs from all other labyrinth fishes by the absence of mouth and throat teeth. They are one of the largest and most specialized of all the labyrinth fishes. The lips also play an important role in interactions between individual fish. The "kissing," for which these fish are known, consists of two specimens facing each other with protruded lips and pushing against each other lip to lip. This "kissing" behavior is not necessarily an indication of a breeding pair, for nonbreeding fish also engage in this behavior.

Helostoma temminckii Cuvier and Valenciennes, 1831
COMMON NAME: Kissing gourami
SYNONYMS: In 1845, Bleeker described two forms as *Helostoma tambakkan* and *Helostoma oligeacanthum.* An-

other synonym is *Helostoma servus* Kaup, 1863. A xanthistic form of this species was described as *Helostoma rudolfi* by Bellanca in 1968.

LENGTH: Up to 12 in. (30 cm), but usually less, and rarely more than 6 in. (15 cm) in the aquarium.

FIN RAY COUNT: D, XVI–XVIII/13–16; A, XIII–XV/17–19; P, 9–11; Pel, I/5; C, 13.

DISTRIBUTION: The kissing gourami is widely distributed throughout Indochina and in most of the Indo-Malayan area. No longer is it possible to determine how much of its distribution is natural and how much is artificial. This species is highly valued as a food fish and has been released in all types of habitats. Adult fish are found in ponds and lakes, but they also occur in running water. With the start of rainfall and floods during the monsoon season the fish wander into the wide flooded lowlands, and they spawn there. They do not take care of their brood. One adult female can lay up to 10,000 eggs in one spawning.

DESCRIPTION: In the aquarium trade the species is represented almost exclusively by a whitish yellow xanthistic form. Wild fish have a greenish gray background color, and their sides show an iridescent greenish or bluish color. A number of light brown horizontal stripes adorn the body. The xanthistic breeding form is unicolored yellowish to a pinkish white.

MAINTENANCE AND BREEDING: The species is well suited for larger community aquariums containing large fishes and small specimens can be kept together with other small fishes. *Helostoma temminckii* is an excellent algae eater. Although in nature they are feeding specialists, in an aquarium they are versatile feeders, greedily devouring dry food or any other kind of food offered.

Since kissing gouramis are sexually mature when they are about 4 in. (10 cm) long, they can be bred in a large aquarium. There are no clearly visible external sex differences. At spawning time the female can be recognized by her increased girth. In the wild form, the males show more intensive colors and dark dorsal and anal fins during the spawning period. In the xanthistic form, however, these changes are too subtle to detect. For breeding purposes only aquariums with a volume of about 25 gallons (200 liters) and upwards should be used. To bring the fish into breeding condition, they should be richly fed with daphnia, adult brine shrimp, soaked oatmeal flakes, plant flakes, and defrosted frozen spinach. During the courtship rituals the fish "kiss" on the mouth and sides. During the embrace the female remains in her usual upright position while the male wraps himself under her. Matings usually take place at dusk and even at night. The eggs, which are lighter than water, rise to the top but often stick on plants because of their adhesiveness. Since these fish eat their own eggs, the parents should be taken out of the aquarium after spawning is complete. The eggs hatch after about 20 hours at a temperature of 79 to 83°F (26 to 28°C). The larvae are free swimming three days later. They are very small and must be fed microscopic size food. Raising a brood requires large amounts of food, and even in large aquariums, only a part of the brood can be raised. Somewhat larger young fish can be fed finely ground dry fish foods or newly hatched brine shrimp nauplii.

Family Osphronemidae

Genus *Osphronemus*

Just as is with the genus *Helostoma,* the monotypic genus *Osphronemus* is also evolutionarily a very old genus. In older literature this fish was often taxonomically placed together with *Trichogaster* and *Colisa* species, although the genus *Osphronemus* is not closely related to them. In *Osphronemus,* and in the genera *Parosphromenus* and *Pseudosphromenus,* the endings *nemus* and *menus* have both been used. The ending *nemus* is probably correct for all. But because in modern aquarium literature the ending *nemus* is generally used for *Osphronemus* and the ending *menus* for *Parasphromenus* and *Pseudophromenus,* these spellings have been kept.

Osphronemus goramy Lacépède, 1802

COMMON NAME: Giant Goramy

SYNONYMS: *Osphronemus goramy* Lacépède, 1802; *Osphronemus olfax* Cuvier, 1817; *Osphronemus satyrus* Bleeker, 1845; *Osphronemus gourami* Regan, 1909.

LENGTH: Up to 24 in. (60 cm), but individual specimens have been known to reach a length of 28 in. (70 cm). This species can be bred at a length of 5 to 7 in. (12 to 14 cm).

These photos show the spawning sequence of *Pseudosphromenus dayi.*

TOP LEFT: Male *Pseudosphromenus dayi* with a bubblenest on the roof of a cave (flower pot).

TOP RIGHT: After spawning in the upper part of the flower pot, the pair sinks to the bottom. The male still holds the female in a tight embrace.

BOTTOM LEFT: The pair has sunk to the bottom. The male releases the embrace, and the female begins to swim away.

BOTTOM RIGHT: Both fish now take the eggs into their mouths in order to carry them to the nest under the roof of the cave.

FIN RAY COUNTS: D, XI–XIII/11–13; A, IX–XII/16–22, Pel, I/5; C, 12–13.

DISTRIBUTION: According to Rachow, this species originally occurred in Java and possibly in Sumatra and Borneo. Today it is distributed throughout Southeast Asia and in warm climates on other continents as well. Of all freshwater fishes, only the carp, and in modern times, the California rainbow trout, probably have been so widely distributed by humans.

Living specimens were first imported into France in 1837 and bred there. Matte was successful in breeding it in Germany in 1877. Later, this fish was bred several times in large aquariums, show aquariums, and similar facilities.

DESCRIPTION: Small young fish an inch or so long are imported now and then and sold in the trade. Such young fish have an elongated snout with a saddle-like indentation over the eyes. Their background color is greenish gray with a number of vertical stripes across the body and a dark spot with a light yellow edge on the caudal peduncle. This juvenile color pattern disappears with growth, and the color becomes gray all over. The dark-edged scales then form a netlike pattern on the gray background. When the fish are about 5 in. (12 cm) long, the fins change to a greenish blue color. Their body form also changes; the body becomes higher, and the elongated head becomes shorter in proportion to the body length. The forehead profile which first was indented now bulges outward and changes more and more into a frontal gibbosity, especially older males. These large goramies are excellent as exhibits for all show aquariums. For the average large home aquarium the giant goramy is suitable only when young.

MAINTENANCE AND BREEDING: In mature fish it is not difficult to distinguish the sexes. The tips of the dorsal and anal fins of the male are elongated and pointed, whereas in the female the tips of these fins are rounded. Varying nest building behaviors have been observed in the giant goramy, both at the surface and below it. Adult females can lay several thousand eggs, which are somewhat oval and have a diameter of 2.5 to 3 mm. The eggs hatch after about 48 hours at a temperature of 77 to 83°F (25 to 28°C); at lower temperatures their development is delayed. The larval fish begin swimming free three to five days later, depending on the temperature, and grow very quickly. Within six to seven months they reach a length of 5 to 6 in. (12 to 15 cm) and are capable of reproduction. The brood-caring males guard their young for some time; this does not happen often in other labyrinth fishes. Giant goramies are very heavy eaters and take much plant as well as animal food. Larger fish eat not only algae but also higher aquatic plants (as a substitute they accept lettuce).

Family Anabantidae

Both the Asiatic genus *Anabas* and the African genera *Ctenopoma* and *Sandelia* belong to the family Anabantidae.

Genus *Anabas*

Anabas testudineus (Bloch, 1795); this is a monotypic genus.

COMMON NAME: Climbing fish

SYNONYMS: *Anthias testudineus* Bloch, 1792; *Lutianus testudo* Lacépède, 1803; *Cojus cobojius* Hamilton and Buchanan, 1822; *Anabas scandens* Cuvier and Valenciennes, 1831; *Anabas oliogolepis* Bleeker, 1854.

LENGTH: Up to 10 in. (25 cm).

FIN RAY COUNT: D, XVI–XIX/7–10; A, IX–XI/8–11; P, 13; C, 13.

DISTRIBUTION: Of all labyrinth fishes, this species has the largest area of distribution. This fish lives on the entire Indian subcontinent, along a line running from Pakistan in the northwest to southern China in the east, and on all the islands of the Malay Archipelago. It also occurs on the Moluccas, the Philippines, and on Taiwan.

Some occurrences probably go back to releases by humans. This is certain for the occurrences on the Celebes, the Philippines, the Moluccas, and on Taiwan, but is also probable for some other places. The Asian climbing fish lives in every type of aquatic habitat, including brackish water. It has a very well-formed labyrinth organ, so that it can leave the water for several hours when water conditions become unfavorable. If necessary it can migrate overland to other bodies of water. Experiments have shown that climbing perch fare quite well on land, as long as they remain in moist grassy areas. It is not true, however, that they eat worms and snails during their land migra-

TOP: *Ctenopoma kingsleyae,* a species widely distributed in Africa. Specimen is not fully grown; adult fish of this species have a heavier appearance and a more rounded snout.
BOTTOM LEFT: An adult male of *Ctenopoma acutirostre* 6 in. (15 cm) long.
BOTTOM RIGHT: The young fish of *Ctenopoma acutirostre* have beautiful markings.

tions. As far as has been observed, they take their food exclusively in the water. In their native waters they are said to wander into the flooded rice fields and eat the grains of rice; aquarium experiments have shown that they do indeed eat grains of rice. They are omnivorus fish that eat a variety of meaty and vegetable foods. Their behavior, however, is more reminiscent of that of predatory fish.

Climbing perch are of great significance as food fish in all of Southeast Asia. Because of their tenacity, they can be brought to market alive; in a tropical climate this is an important advantage, because they don't spoil as rapidly as dead fish do. They are caught with nets and traps, with rods, and dug from the moist bottom mud of evaporated ponds. This fish was first imported in 1891 and first bred shortly thereafter.

DESCRIPTION: *Anabas* has gill covers that are equipped with numerous spines, which aid their forward motion during land migrations. These spines are also significant as defensive weapons. Climbing fish often get stuck in a net because of these spikes.

Their body is elongated and compressed slightly on the sides. The background color is olive to gray, but grayish blue also occurs. In some localities they have very light vertical stripes, especially younger fish. The young fish also have a dark spot on the caudal peduncle most of the time; the spot disappears with increasing age. There are no color differences between males and females, but the sex can easily be determined by the shape of the body. The females are more stocky and have a larger girth than the males, especially during the spawning period.

MAINTENANCE AND BREEDING Because of their predatory nature, these fish are not suitable for community aquariums. At best they can be kept together with rather large species of *Ctenopoma* or with cichlids of about the same size. These fish are therefore kept as an aquarium fish only by dedicated specialists. Because of its peculiarities, the climbing perch is, however, very suitable as a showfish for public aquariums. A large aquarium is required for breeding this fish. Spawning usually occurs near the bottom, but no territory is staked out and no nest is built. During the embrace the female is usually only partially turned over. The clear eggs released during spawning rise to the surface. They are quite small, having a diameter of only 0.5 mm. Full-grown females can lay 2000 to 4000 eggs. The eggs hatch in about 24 hours at a temperature of 77 to 83°F (25 to 28°C). About three days later the tiny young fish swim free. At first they must be fed microorganisms such as paramecium. There is no parental brooding. The climbing fish is most active at dusk or at night, and that is usually when spawning begins.

African Labyrinth fishes

Family Anabantidae

Genus *Ctenopoma*

The genus *Ctenopoma,* established by Peters in 1844, contains about 25 species, several of which are probably duplicate descriptions; thus the actual number of species is probably smaller. Their distribution is limited to tropical Africa. Some species show certain common traits with the Asiatic climbing perch (*Anabas testudineus*) and like the climbing perch, all species have serrated gill covers. Some of them are also able to migrate overland.

Ctenopoma species occur in all large river systems south of the Sahara. They are present not only in the extreme southern part of the African continent, but they are also found in bays and quiet sections of large rivers, such as the Niger, Congo, Zambesi, and others, as well as in their tributaries. *Ctenopoma* species also occur in smaller bodies of water such as ponds and marshy lakes. Certain species temporarily live in flooded areas during the rainy season. The larger species are of economic importance; as food fish they play a role similar to that of their relatives in Southeast Asia. For several reasons *Ctenopoma* are rarely encountered in the aquarium trade. One is that far fewer fish are imported from Africa than from Southeast Asia. Another is that they are usually less conspicuously colored than the Southeast Asiatic labyrinth fishes (most hobbyists prefer brightly colored fishes). In addition, they are frequently very shy in an aquarium, and many are active only nocturnally. Some are also quite predatory and can be kept only in special aquariums. Some of the smaller *Ctenopoma* species, however, are suitable as aquarium fishes and can be bred.

Overall, fishes of the genus *Ctenopoma* are quite difficult to classify, as they are quite variable in many of their characteristics. A thorough systematic revision of this genus is desirable, but our knowledge of these fishes is fragmentary, and a revision of the genus in the near future is probably not to be expected.

There are some *Ctenopoma* species that offer no brood care. They do not even establish territories. In the aquarium these species are extremely peaceful toward each other. Males have a number of special scales equipped with spines. These spiny scales are located on each side of the head behind the eye and on the caudal peduncle. In some species there are two groups of spines at each location. For a long time the purpose of these spines was not clear, but by close observation of the mating behavior of these fishes (see species description for *C. muriei*) it was determined that the function of these spines is to help hold the female securely during these fishes' rapid embrace. The spines thus have a purpose similar to that of the barb on the anal fin of certain male characoids, which are far more familiar to aquarium enthusiasts. The courtship of these free-spawning *Ctenopoma* species is also similar to that of the characoids. The males are quite aggressive prior to mating, and often a number of fish simultaneously participate in the spawning. The embrace occurs very rapidly. The eggs released during the embrace rise to the surface. Floating larvae emerge from the eggs, and they continue to float on the surface until absorption of the yolk sac is complete.

There are some species with only one group of spines behind each eye. These fishes are slender, active, and, at least in an aquarium, very quarrelsome. To date, no details are known about the reproduction of these species. It can be assumed that they, too, are non-brooding fishes. The better-known species, (e.g., *C. acutirostre, C. kingsleyae, C. muriei* and *C. oxyrhynchus*) have two spine groups behind each eye. *Ctenopoma multispinis* and *C. pellegrini,* on the other hand, have only one. Scale spines are a reliable sexual characteristic of males, in those *Ctenopoma* species that have them. Sometimes old females show beginnings of scale spine formation. The fact that overage and probably no longer fertile females develop secondary sexual characteristics of males is not unusual. It occurs, for example, with the frontal gibbosity developed in old females of the African cichlid *Haplochromis moorii.*

Some *Ctenopoma* species are broodcaring bubblenest builders whose broodcare behavior is similar to that of the bubblenest building Asiatic labyrinth fishes. As of this writing, only a few such species have reproduced regularly in an aquarium, and most of the important details of their spawning and broodcare behavior are known. A deviation from the Asiatic bubblenest builders occurs in the position of the female during the embrace. She usually maintains her normal right-side-up position. In at least one species the males guard their nests only from a distance.

Taxonomically, among the species with two spine groups behind each eye *C. petherici, C. kingsleyae* and *C. argentoventer* represent a very closely related species complex. A few other species may also belong to this complex. These fish probably represent one species with different subspecies distributed over large parts of Africa. Because a more thorough investigation of the relationships among these species has not yet been carried out, the customary names for these fishes in aquarium literature will be used in the species descriptions. If more detailed investigations should show that these fishes belong to a single species, then, according to the rule of priority, the name *Ctenopoma petherici* would be valid.

In some older publications it has been stated that there are also mouthbrooders among the *Ctenopoma* species. This belief dates back to a note by G. A. Boulenger that he had found a number of eggs in the gill cavity of a female *C. multispinis.* In 1971 an investigation by Peters showed that Boulenger's preserved specimen had no eggs but had the cysts of a parasite (*Myxobalus* sp.) in its gill cavities.

Large *Ctenopoma* species should not be kept together with small species. It must be remembered that predatory species, such as *C. acutirostre* and *C. ocellatum,* eat any smaller fishes without hesitation, but fishes too large to be eaten will be left alone. These predatory species should be kept in separate aquariums. Some of the small, timid species, such as *C. nanum* and *C. damasi,* are also best kept by themselves, especially if one wants to breed them.

All *Ctenopoma* species are heavy eaters, and require very nourishing food in order to thrive in aquariums. In addition to daphnia and mosquito larvae, larger species can be fed earthworms and small feeder fish. Most individuals accept commercial dry food in flake form.

Ctenopoma species also have spines on their gill covers, and when caught with a net, they frequently get stuck. Usually the net has to be cut to free the fish. Hobbyists who keep these fishes should use a large glass jar to trap them.

Broodcaring *Ctenopoma* species

Ctenopoma ansorgei (Boulenger, 1912)
COMMON NAME: Orange Ctenopoma
SYNONYM: *Anabas ansorgei* Boulenger, 1912.
LENGTH: Males up to 3 in. (8 cm), females smaller.
FIN RAY AND SCALE COUNTS: D, XVII–XVIII/7; A, X–XI/7; P, 7; Pel, I/5; C, 12; L.l. scales, 28–30.
DISTRIBUTION: This fish is known from Cameroon and the lower Congo River, including Stanley Pool. The species seems to occur mainly in the smaller tributaries of larger rivers. Linke found these fish near Kribi in southern Cameroon, in running water under embankments in which there were many hiding places. The water was acidic, and the carbonate hardness was less than 1 DH. Orange ctenopoma appeared in Holland in 1957. The importer is not known to the author. Offspring from these Dutch imports have been on the market since 1958.
DESCRIPTION: The body is elongate, and the head is uniformly tapered anteriorly. This fish is one of the few truly colorful *Ctenopoma* species. Normally it shows a light brownish yellow background color over which run six darker vertical stripes. At spawning time these fish change color, but especially the males; their ver-

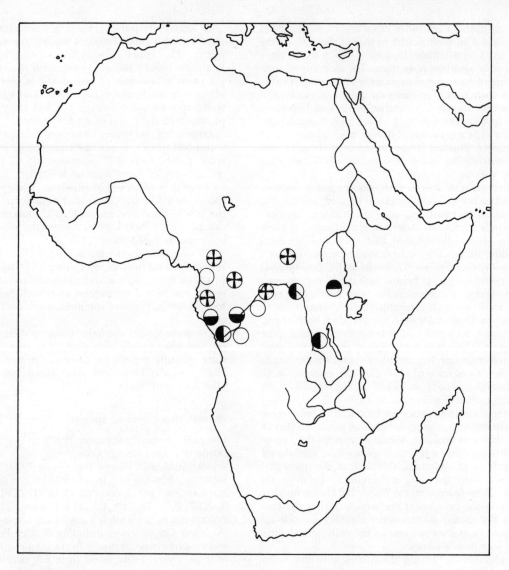

Distribution map of the broodcaring species of the genus *Ctenopoma*

○ *Ctenopoma ansorgii*

◐ *Ctenopoma congicum*

◑ *Ctenopoma damasi*

◖ *Ctenopoma fasciolatum*

⊕ *Ctenopoma nanum*

tical stripes fade and are replaced by greenish iridescent stripes. The dorsal and anal fins become orange with diagonal dark brown to black stripes. The dorsal and anal fin stripes are a continuation of the vertical stripes that run across the body. These colors are duller in females. Adult males frequently have elongated soft rays on the posterior end of the dorsal and anal fins. The caudal fin is brown to reddish brown, and the elongated pelvic fins are white in front and orange at the base.

MAINTENANCE AND BREEDING: *Ctenopoma ansorgei* are best kept by themselves. They are peaceful toward other fishes, but because of their shy nature they are easily intimidated, and in the presence of other fishes they remain hidden most of the time and will not compete for food. Soft water should be used for breeding. The breeding pair should be well supplied with live food, if possible with mosquito larvae and brine shrimp. This species is a bubblenest builder, and its spawning behavior is similar to that of typical nest-building Asiatic labyrinth fishes. Most of the time, however, the female is not turned upside down during the embrace. According to H. J. Richter, up to 6000 eggs are produced in one spawning. The eggs hatch in 24 hours at 77 to 79°F (25–26°C), and three to four days later the larval fish are free swimming. At first they must be fed very small food organisms as paramecium. The young fish grow quickly and show signs of their adult coloration in only three weeks.

Ctenopoma congicum Boulenger, 1887
COMMON NAME: Congo ctenopoma
SYNONYM: *Anabas congicus* Boulenger, 1899.
LENGTH: Males up to 3 in. (8 cm), females smaller.
FIN RAY AND SCALE COUNT: D, XVI–XVII/8–9; A, IX–XI/9–11; L.l. scales, 26–28.
DISTRIBUTION: Gabon and the lower basin of the Congo River. This fish was first imported in 1975.
DESCRIPTION: The body shape is similar to that of *C. ansorgeii,* but the head is more pointed. The background color is yellowish brown to reddish brown. At times seven to eight dark vertical stripes are present. There are a number of darker spots in the dorsal fin, and six to seven dark vertical stripes in the anal fin. In adult males, the dorsal and anal fin tips are pointed, and they are rounded in females. This characteristic pronounced in all individuals, especially in younger ones. Most of the time it is simpler to distinguish the sexes by the body form. The female is stouter.
MAINTENANCE AND BREEDING: This species should be kept in soft water. These fish show a preference for low light intensity. The males build their bubblenests in the darkest locations of the aquarium, preferably

under a thick clump of floating plants. The development of eggs and larval fish does not differ from that of *C. ansorgii.*

Ctenopoma damasi (Poll, 1939)
COMMON NAME: Pearl ctenopoma
SYNONYM: *Anabas damasi* Poll, 1939.
LENGTH: Males up to 2¾ in. (7 cm); females up to 2½ in. (6 cm).
FIN RAY AND SCALE COUNT: D, XIV–XIX/6–9; A, IX–XII/6–9; L.l. scales, 26–30.
DISTRIBUTION: Uganda and parts of Zaire bordering Uganda. According to Greenwood (1958) and Poll (1939) the pearl ctenopoma occurs in small ponds and marshes in the catchment basin of Lake Edward. Pearl ctenopoma were first imported in 1968 by Berns and Peters, of the Zoophysiological Institute of Tübingen University.
DESCRIPTION: The body is elongated, the head is large, and the background color in the male is blackish gray. During spawning periods this color becomes an iridescent blackish blue on a dark background, embellished by numerous iridescent blue-green spots. Broodcaring males in particular show intensive coloration. The females are a lighter gray-blue, and their spots are indistinct.
MAINTENANCE AND BREEDING: According to Berns and Peters, the males build small firm bubblenests. Spawning behavior does not differ from that described for *C. ansorgei.* After spawning is completed, the males guard the nest from some distance, while lying on the bottom of the aquarium, a somewhat unusual behavior in bubblenest-building labyrinth fishes. The eggs hatch after 36 hours at 80°F (27°C), and three to four days later the larval fish are free swimming. They are about 2mm long and must initially be fed small microorganisms. Four to five days later they can manage brine shrimp nauplii.

According to Richter, females may change partners during spawning, a behavior that also occurs in other bubblenest-building labyrinth fishes, but which will occur only in very large aquariums.

Ctenopoma fasciolatum (Boulenger, 1899)
COMMON NAME: Striped ctenopoma
SYNONYMS: *Anabas fasciolatus* Boulenger, 1899; *Ctenopoma fasciolatus* Myers, 1926.
LENGTH: Males up to 3½ in. (8 to 9 cm); females up to 2¾ in. (7 cm).
FIN RAY AND SCALE COUNT: D, XVI/9–11; A, X/9–11; L.l. scales, 27–28.

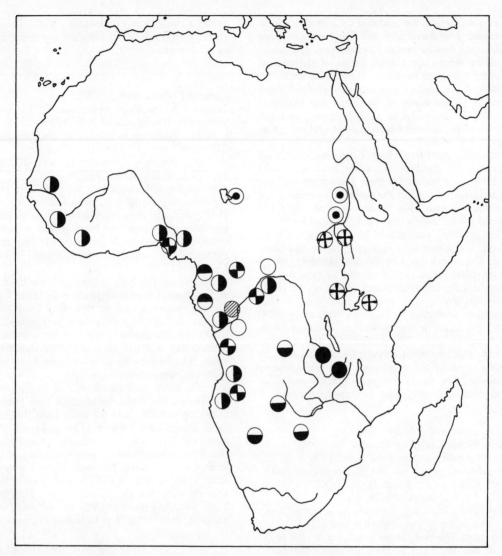

Distribution map of the nonbroodcaring species of the genus *Ctenopoma*

○ *Ctenopoma acutirostre*　　　　⊕ *Ctenopoma muriei*　　　　⊙ *Ctenopoma petherici*

◑ *Ctenopoma argentoventer*　　　◓ *Ctenopoma oxyrhynchum*

◐ *Ctenopoma kingsleyae*　　　　◔ *Ctenopoma nigropannosum*

⊖ *Ctenopoma maculatum*　　　　● *Ctenopoma ocellatum*

⊖ *Ctenopoma multispinis*　　　　▨ *Ctenopoma pellegrini*

DISTRIBUTION: The striped ctenopoma is found throughout the entire Congo River basin. It lives mostly in deep, clear water. Within its large geographical range, there are several color types. Within these types is a range of background colors varying from grayish blue to yellowish brown. Eight to nine lighter vertical stripes run across the body, but the stripes are not equally visible in all forms. The fish was first imported in 1912.

DESCRIPTION: The body form is medium high with laterally compressed sides. The background color is yellow to yellowish brown with eight to ten dark brown vertical stripes. These stripes can be almost black when the fish are healthy, but are obscure when the fish are stressed, for example, when the temperature drops, or when the pH is too high or too low. The males have a blue iridescent spot on the gill covers and long pointed dorsal and anal fins. The paired fins show a fine pattern of lines and spots on a light yellowish brown background. In an aquarium they are quite lively—not nearly as shy as most other *Ctenopoma* species, and they are more diurnal than most of their relatives.

MAINTENANCE AND BREEDING: Of all the *Ctenopoma* species this one is best suited for a community aquarium. The fish are peaceful toward other fish and toward each other. They do not demand special water conditions, and they accept almost any kind of food. They are relatively easy to spawn. The male builds his bubblenest under a thick layer of floating ferns or similar plants. Since these fish are quite fecund, a large breeding aquarium should be used. During the spawning embrace the male curls under the female while the female retains her normal upright position. The eggs hatch in about 35 hours at a temperature of 76 to 79°F (24 to 26°C). The young fish are free swimming three to four days later and must at first be fed small microorganisms. Broodcaring males of this species can be very aggressive. In smaller breeding aquariums, the female should be removed right after spawning. Caution should also be taken to protect the other fishes in a community aquarium when the male starts building a bubblenest.

Ctenopoma nanum Günther, 1896

COMMON NAME: Dwarf ctenopoma.

SYNONYMS: *Anabas maculatus* Boulenger, 1882; *Anabas nanus* Boulenger, 1916.

LENGTH: Males to 2¾ in. (7 cm); females to 2½ in. (6 cm).

FIN RAY AND SCALE COUNT: D, XV–XVII/7–10; A, VII–IX/9–11; C, 12; L.l. scales, 25–30.

DISTRIBUTION: This species is known from southern Cameroon, from Gabon and Zaïre, where *C. nanum* is found in the tributaries of the Congo, Ubangi, Uee Rivers and in Stanley Pool.

DESCRIPTION: *Ctenopoma nanum* is similar in appearance to *C. fasciolatum* but is more slender. The background color is brownish yellow with seven or eight dark vertical stripes that extend into the dorsal fin. These vertical stripes are particularly visible in young fish and in females, but they are faint in adult males. The males change color during the spawning period, turning dark gray to blue-gray and sometimes almost black. The dorsal and anal fins assume a bluish hue against which the vertical stripes (which have completely disappeared from the body) clearly contrast. The vertical stripes in the females are interrupted during the spawning period by a wide yellowish gray horizontal stripe that runs from the gill covers to the caudal peduncle. The females are easily distinguished from the males even in the nonspawning period, for they are smaller and have rounded tips on the dorsal and anal fins. These fin tips are pointed in the male and in older specimens are somewht extended out.

MAINTENANCE AND BREEDING: When fed live food, especially mosquito larvae, and when kept in soft, peat-filtered water, dwarf ctenopoma are relatively easy to breed. Although they are not very particular about water conditions, they spawn more readily in soft water. The males build their bubblenests under pads of floating plants, under large floating leaves, or in a similarly concealed place. Spawning usually begins before the bubblenest is finished. Usually it begins when the female spontaneously approaches the male. The female shows her readiness by pushing her mouth into the male's side. That is the same type of behavior seen in most of the Asiatic labyrinth fishes. In the embrace the male lies curled under the female. The eggs are not released until the slowly sinking pair drifts to the bottom. Thirty to 50 eggs are released in each spawning and rise to the nest where they are supported with new bubbles made by the male. Before the male begins this job he first chases the female away. The male of the dwarf ctenopoma is rather aggressive toward the female. In one spawning a total of about 300 to 500 eggs are laid. According to Richter, there can be up to 1000 eggs in one spawning. The eggs hatch in 24 hours at a temperature of 77 to 79°F (25 to 26°C). Development does not differ from that described for *C. ansorgei*. Dwarf ctenopoma are shy fish that are not well suited for the community aquarium.

Nonbroodcaring *Ctenopoma* species

Ctenopoma acutirostre Pellegrin, 1899

COMMON NAME: Leopard ctenopoma

SYNONYMS: *Ctenopoma petherici* Schilthuis, 1891; *Anabas weeksii* Boulenger, 1905.

LENGTH: Up to 8 in. (20 cm), rarely reaching this length in an aquarium.

FIN RAY AND SCALE COUNT: D, XIV–XVIII/9–12; A, IX–X/10–12; P, 14; Pel, I/5; C, 14; L.l. scales, 26–28.

DISTRIBUTION: The lower and central river basin of the Congo. First imported in 1956.

DESCRIPTION: The leopard ctenopoma is a typical predatory fish, which is most active at dusk or at night. Under suitable cover the fish waits, completely motionless, in ambush for its prey, which it grabs with a quick, short lunge. Only younger specimens pursue their prey.

The body is high and laterally compressed. Since the dorsal and anal fins have the same marking pattern as the sides of the body, and their bases are covered with small scales, their body appears to be higher than it really is. The posterior part of the body merges directly into the caudal fin without a tail trunk much of a caudal peduncle, and the caudal peduncle is also covered at the base with small scales. The soft parts of the dorsal and anal fins extend all the way to the caudal fin, thus appearing to be one continuous fin. The snout is pointed and the head is large in relation to the body length. The mouth is deeply cleft and can be protruded forward. The eyes are noticeably large. The sexes can only be recognized by the spine groups of the males.

The background color is light yellowish brown with an irregular dark brown spotted pattern. These spots can temporarily disappear in older fish and are replaced by a uniform dark brown color.

No details are known about their breeding behavior. This fish has reproduced in public aquariums. The young fish show an interesting spot pattern.

Ctenopoma argentoventer (Ahl, 1922)

SYNONYMS: *Anabas argentoventer* Ahl, 1922; *Ctenopoma argentoventer* Myers, 1924.

LENGTH: Up to 6 in. (15 cm).

FIN RAY AND SCALE COUNT: D, XVI/10; A, IX/10; L.l. scales, 26.

DISTRIBUTION: According to Ahl this species is found in the Niger River delta and the lower Niger River. It was first imported in 1912. Arnold introduced it as *Anabas* spec. in the journal *Blätter*. This species was later imported on several occasions, together with other West African fishes.

DESCRIPTION: The body is long. The head is uniformly tapered to a point. The color is variable and depends on mood, changing from yellowish gray almost to black. The ventral side is gray to silver. Beginning at the eye, a yellowish stripe curves toward the front part of the back. A second stripe runs along the center of the body, at about the height of the front of the anal fin base. It is uncertain whether or not these two stripes are a sex-distinguishing characteristic. I saw about 40

young fish probably of this species at a Belgian import company in 1956, and all of the fish showed these stripes. Unfortunately their exact place of origin could not be determined. The stripes disappear with advancing age. In contrast, older fish keep a lightly edged dark spot on the caudal peduncle.

MAINTENANCE AND BREEDING: Some time ago this fish was bred in captivity. Breeding does not differ from that described under *C. kingsleyae*. In their book *Exotic Freshwaterfishes* (1936) Arnold and Ahl repeat a statement by Schreitmüller, who said that a cross between *C. argentoventer* and *Anabas testudineus* was successful in an aquarium at the zoo in Frankfurt/Main in 1919, but this statement is unconfirmed and has to be considered doubtful.

The species status of *C. argentoventer* is uncertain (cf. remarks under *C. kingsleyae*).

Ctenopoma kingsleyae Günther, 1896

SYNONYM: *Anabas kingsleyae* Boulenger, 1899.

LENGTH: To 8 in. (20 cm).

FIN RAY AND SCALE COUNT: D, XVI–XVII/8–10; A, IX–X/9–11; L.l. scales, 25–29.

DISTRIBUTION: The species is widely distributed over all of West Africa, from Gambia in the north to Angola in the south. It also occurs frequently in the Congo River basin up to Stanley Pool. *Ctenopoma kingsleyae* is found in a wide variety of habitats, but mainly in running water. In the northern part of its distribution it often enters wide, flooded areas during the rainy season, where it spawns. As the water level drops the fish often remain trapped in small ponds. If the ponds dry out, the fish reach other water by migrating overland. They can also survive shorter dry periods buried in moist bottom mud. Of the closely related species *C. kingsleyae* and *C. petherici*, the former has

TOP LEFT: Croaking dwarf gourami (*Trichopsis pumilus*).

TOP RIGHT: Schaller's croaking gourami (*Trichopsis schalleri*).

CENTER LEFT: Croaking gourami (*Trichopsis vittatus*), male.

CENTER RIGHT: Climbing perch (*Anabas testudineus*).

BOTTOM LEFT: Orange ctenopoma (*Ctenopoma ansorgii*) is one of the smallest *Ctenopoma* species. It is a broodcaring bubblenest builder.

BOTTOM RIGHT: Congo ctenopoma (*Ctenopoma congicum*), male. This species also cares for its brood.

the much wider distribution. If it should turn out that both belong to one species, *C. petherici* is to be considered the nominate species.

DESCRIPTION: The body is elongated and almost uniformly oval in a cross-section. The fish is uniformly tapered toward the snout. The color depends strongly on the fish's physical condition, but it is also variable within the large geographical range of the species. The color varies from light yellowish gray through bluish gray and brown, almost to black. The scales have dark edges that stand out clearly, especially in adult specimens. Mature females are especially noticeable by the fullness of their bodies. A definite sexual characteristic of the males is their spine groups, which, in this species are not always visible under every light. They are best recognized when one places the fish in a small glass container and observes it under reflected light.

MAINTENANCE AND BREEDING: *Ctenopoma kingsleyae* is very undemanding and insensitive to different environments. Because of its size it is not recommended for community aquariums. It can be kept in very large aquariums with fish of equal size. They are peaceful toward each other. Like other large *Ctenopoma* species, these fish develop slowly and do not mature prior to their second or third year of life. On the other hand, they can live to a very old age, 20 years not being uncommon.

They are not particular about their food, but spawn only if properly fed. Foods that bring on breeding are mosquito larvae, earthworms, and small feeder fish. The parents do not care for their brood, often devouring their eggs right after spawning. Spawning is preceded by a lively courtship and the embraces occur with lightning speed. Thousands of eggs are released during one spawning. They rise to the surface and remain there until they hatch. The rate of the embryonic and larval development as well as that of the young fish strongly depends on the temperature. At a temperature of 85°F (30°C) the eggs hatch in about 24 hours, at 75°F (24°C) they hatch in 48 hours, and at 68°F (20°C) after 96 hours. The fact that the eggs

TOP: A yellowish pink (xanthistic) specimen of the kissing gourami (*Helostoma temminckii*). This form is almost exclusively the only one seen in the aquarium trade.
BOTTOM LEFT: The wild form of the kissing gourami (*Helostoma temminckii*).
BOTTOM RIGHT: Male of *Ctenopoma multispinis* with an easily visible group of scale spines behind the eye.

develop at a temperature as low as 75°F speaks well for the adaptability of this species. The times given were determined in an experiment in which the eggs of one spawning were allowed to develop at different temperatures. Eggs no longer developed at a temperature of 65°F (18°C). It is interesting to note that development of embryos and larvae into young fish was also delayed at temperatures above 30°C. Thus the optimum temperature for development of eggs and larvae seems to lie between 83 and 85°F (28 to 30°C).

The fish are free swimming in two to five days, also depending on the temperature. They can be fed brine shrimp right away. Initially the young fish grow rapidly, but after they reach a length of about 4 cm, their growth slows down.

Ctenopoma maculatum Thominot, 1886.
SYNONYMS: *Ctenopoma weeksii* Boulenger, 1896; *Anabas oxyrhynchus* Steindachner, 1913.
LENGTH: To 8 in. (20 cm).
FIN RAY AND SCALE COUNT: D, XIV–XVI/9–11; A, VII–IX/9–11; L.l. scales, 26–29.
DISTRIBUTION: Southern Cameroon and the Congo.
DESCRIPTION: This species is similar in body shape to *C. kingsleyae* but seems to be more compact. The top of the head is more strongly curved than the bottom. The background color is yellowish brown to brown. A dark black spot lies about in the center of the body, directly under the midline. Young fish exhibit a number of dark vertical stripes, which, at times, are also visible in older fish. It is imported only singly.
MAINTENANCE AND BREEDING: Nothing is known about breeding it. This species is probably a nonbroodcaring *Ctenopoma* species.

Ctenopoma multispinis Peters, 1844
SYNONYMS: *Ctenopoma multispinis* Günther, 1861; *Anabas multispinis* Boulenger, 1905; *Sprirobranchus multispinis* Woosnam, 1910.
LENGTH: To 6 in. (15 cm).
FIN RAY AND SCALE COUNT: D, XVII–XVIII/8–9; A, VIII–X/8–9; L.l. scales, 31–35.
DISTRIBUTION: From southern Zaïre to Botswana and northern Namibia, and in Zimbabwe and Mozambique. It lives at times in flooded lowlands, marshes and similar biotopes and can survive buried in bottom mud for long dry periods.

It is not known when the first importation of this species occurred. Individual specimens were exhibited in public aquariums in the 1930s. Schaller was able to import a large number in 1973, and Foersch bred them.

DESCRIPTION: It is a slender species with only frontal scale spines. The skin is coarse and the scales are large, an adaptation to extreme living conditions such as overland migrations.

The background color is olive yellow to brown. In older specimens the color can temporarily darken. In normal coloration they show a brown spotted marking on a yellowish brown background. Males are distinguished from females by presence of scale spines.

MAINTENANCE AND BREEDING: This species should be kept only with other robust species or in a single-species aquarium. Small fishes in addition to earthworms and mosquito larvae are readily eaten. Reproductive behavior is similar to that of *C. muriei*.

Ctenopoma muriei (Boulenger, 1906)
COMMON NAME: Nile ctenopoma.
SYNONYMS: *Anabas muriei* Boulenger, 1906; *Anabas houyi* Ahl, 1927.
LENGTH: Up to 3¼ in. (8.5 cm).
FIN RAY AND SCALE COUNT: D, XIV–XV/7–10; A,
DISTRIBUTION: Upper White Nile. It also occurs in numerous East African lakes and their drainage areas, for example in the area of Lake Victoria, Lake Albert, Lake Edward, and Lake Tanganyika. It also resides in marshy and at times flooded areas and is found in the smallest bodies of water during the dry periods. It was first imported by Berns and Peters of the Zoophysiological Institute of the University of Tübingen, in 1968.
DESCRIPTION: In spite of considerable size differences, there are certain similarities between this species and *C. kingsleyae*. *Ctenopoma muriei* has a more brownish background color and numerous dark spots on the upper side. The spot on the caudal peduncle is less well-defined than that of *C. kingsleyae* but is still easy to recognize. In addition to the lack of scale spines, females differ from males during nonspawning periods by their duller coloration.
MAINTENANCE AND BREEDING: This species is quite active and should be kept only in a single-species aquarium.

According to Mörike (1977) this species does not take care of its brood. Spawning takes place after much dusk or night activity. The embrace occurs when the female slows down. She stays in her normal upright body position. Spawning only takes two to three seconds and 10 to 30 eggs are released. Since this species spawns in swarms, it is assumed that they change partners now and then during spawning activity. Each female lays 200 to 2000 eggs. The eggs have a diameter of 0.85 mm. The eggs hatch in 22 to 26 hours at a temperature of 80°F (27°C).

Ctenopoma nigropannosum Reichenow, 1875
SYNONYMS: *Ctenopoma gabonese* Günther, 1896; *Anabas nigropannosus*, Boulenger, 1899.
LENGTH: Up to 6¾ in. (17 cm).
FIN RAY AND SCALE COUNT: D, XIX–XX/9–10; A, IX–XI/9–10; L.l. scales, 30–33.
DISTRIBUTION: From the delta of the Niger River to Angola and in the lower Congo and Stanley Pool. It is rarely imported.
DESCRIPTION: The body is elongated and the snout is pointed. The background color is brownish yellow to yellowish gray. Young fish have darker, irregular vertical stripes, which nearly disappear in older fish. There is a dark spot on the gill cover and on the caudal peduncle.

This species is related to the slender and active forms in which males have only frontal scale spines. It has the reputation of being an aggressive species.
MAINTENANCE AND BREEDING: It should be kept only in a single-species aquarium. Nothing is known about breeding it.

Ctenopoma ocellatum Pellegrin, 1899
COMMON NAME: Eyespot ctenopoma
SYNONYM: *Anabas ocellatus* Boulenger, 1905.
LENGTH: The original description gives 5½ in. (14 cm), but much larger specimens are known.
FIN RAY AND SCALE COUNT: D, XVI–XVIII/9–12; A, IX–X/10–12; P, 14–15; Pel, I/5; C, 17; L.l. scales, 26–28.
DISTRIBUTION: Tributaries of the upper Congo.
DESCRIPTION: The eyespot ctenopoma strongly resembles *C. acutirostre*, but it has a similarly pointed head only as a young fish. Older specimens have a distinctively more compressed head. The background color is yellowish brown with irregular vertical stripes, which sometimes break up into individual spots on the anterior part of the body. The arrangement of dorsal, tail and anal fins also reminds one of *C. acutirostre;* looked at superficially, these fins appear as one closed fin edge. There is a dark spot with a light edge on the caudal peduncle.
MAINTENANCE AND BREEDING: This species most likely does not belong to the broodcaring species. The males have scale spines. Nothing is known of its reproduction.

Ctenopoma oxyrhynchus (Boulenger, 1902)
COMMON NAME: Peacock-eye ctenopoma
SYNONYM: *Anabas oxyrhynchus* Boulenger, 1902.
LENGTH: Up to 4½ in. (11 cm).
FIN RAY AND SCALE COUNT: D, XV/10; A, VII/10; L.l. scales, 28.

DISTRIBUTION: Lower Congo, Stanley Pool, and Ubangi. This species was first imported by a Belgian import company in 1951.

DESCRIPTION: It is a high-backed, laterally compressed species. The lower outline of the head profile is almost straight, the upper outline is strongly curved. Younger specimens have a more pointed head. The background color is yellowish brown, with a larger dark spot on the center of the body. The anal and caudal fins are dark-edged. At spawning time the fish takes on a marbeled pattern of irregular black patches on a light brown background. The males are recognizable by their more slender body in addition to their scale spines.

MAINTENANCE AND BREEDING: Since the males are aggressive, large aquariums are recommended for breeding them. The embrace occurs when the female stops her active swimming. Then the fish circle each other, and shortly the embrace occurs, which lasts only a few seconds. According to Richter, it usually occurs near the bottom of the aquarium. The released eggs rise directly to the surface. They hatch in 24 hours, and after another 3 days the young fish are free swimming. They should be fed initially with paramecium or the finest dry natural food. The young fish remain mostly in dark places in the aquarium or under plant leaves where they may cluster in a narrow space.

Ctenopoma pellegrini (Boulenger, 1902)
SYNONYM: *Anabas pellegrini,* 1909.
LENGTH: Up to 4 in. (10.5 cm).
FIN RAY AND SCALE COUNT: D, XVIII–XIX/10–11; A, VII/10; L.l. scales, 33–34.
DISTRIBUTION: The only occurrences are from the Ubangi River area.
DESCRIPTION: Together with *C. multispinis* and *C. nigropannosum,* this species belongs to the slender forms in which the males have frontal scale spines. No observations about keeping it in an aquarium are available.

Ctenopoma petherici Günther, 1864.
SYNONYM: *Anabas petherici* Boulenger, 1899.
LENGTH: Up to 6¼ in. (16 cm).
FIN RAY AND SCALE COUNT: D, XVII–XIX/8–10; A, X/10; L.l. scales, 28–30.
DISTRIBUTION: This species occurs in the Sudan, in the area of the tributaries of the White Nile, Jur, and Bahr el Djebel. Additional occurrences are known from Lake Chad and its vicinity. It is rarely imported.
DESCRIPTION: This species is extremely close taxonomically to *C. kingsleyae.* It is possible that both are merely geographical forms of the same species.

Genus *Sandelia*

This genus, established by Castelnau in 1861, contains only two species. They are limited in their distribution to the Cape Province in the extreme southern part of the African continent. There is no geographical connection between them and the *Ctenopoma* species. The closest occurrences of *Ctenopoma* in Botswana and northern Namibia are separated from the occurrences of the *Sandelia* species by far more than 620 mi (1000 km). These are elongated fishes with a body almost oval in cross section. The head is large and the mouth is deeply cleft. The dorsal fin is long with numerous spinous rays, and the anal fin is about half the length of the dorsal fin. The caudal fin is rounded and the pectoral fins begin far down the side in the lower third of the body. The gill cover scales have smooth edges, except for two small spines. The lateral line is interrupted and continues two scale rows lower in the rear part of the body. The labyrinth organ is small and not very convoluted.

These fish seem to deviate from all other labyrinth fishes in their reproductive behavior. As far as is known, they have not yet bred in captivity, but there are breeding observations in nature for *S. capensis* by Jubb in 1967. According to Jubb's report, this species becomes mature in its first year of life. The fish spawn in spring and mature in summer in shallow quiet waters, but they do not build a bubblenest. Their eggs are reported to sink to the bottom and stick to plants and other objects. After spawning, the male stays at the spawning site and guards the eggs. The eggs have a diameter of about 1 mm. They hatch after 35 hours at a temperature of 75°F (24°C), and the young fish are free swimming three days later.

In some reports a third species of *Sandelia* is listed under the name *Sandelia vicina* (Boulenger, 1916). The status of this species is doubtful; it may be a local form of *S. capensis.*

Sandelia bainsii Castelnau, 1861
SYNONYMS: *Ctenopoma microlepidotum* Günther, 1861; *Spirobranchus bainsii* Günther, 1861; *Anabas bainsii* Boulenger, 1905.
LENGTH: Up to 7 in. (18 cm).
FIN RAY COUNT: D, XV–XVII/9–10; A, VII–VIII/9–10.
DISTRIBUTION: This species is found only in a small area about 420 mi (700 km) east of Capetown, in the coastal region between the cities of Port Elizabeth and East London. It lives mainly in smaller rivers.
DESCRIPTION: Adult fish have a bluish gray primary background color with the sides lighter gray. Younger specimens have a more yellowish gray background color

and nine to ten dark vertical stripes. Nothing is known about breeding or keeping them in the aquarium, but in the last few years individual live fish have been brought to the West.

Sandelia capensis (Cuvier and Valenciennes, 1831)
SYNONYMS: *Spirobranchus capensis* Cuvier and Valenciennes, 1831; *Anabas capensis* Boulenger, 1905.
LENGTH: Up to 8½ in. (21 cm), usually smaller.
FIN RAY COUNT: D, XII–XIV/8–10; A, VI–VII/8–11.
DISTRIBUTION: This species is limited to the western part of Cape Province where it occurs in the Langevlei River, the Elands River, the Camtoos River, and the Coega River. First live import was of two specimens to Holland in 1958, and again in 1973.
DESCRIPTION: This species is not as slender as *S. bainsii*. The cleft of the mouth is deeper. The background color is olive brown, becoming lighter toward the bottom and changing there to yellowish. According to the type of habitat there are subtle differences in color. Younger specimens have striped and spotted markings of diffuse brown spots.
MAINTENANCE AND BREEDING: This is a predatory fish, suitable only for a single-species aquarium.

Pikeheads

Family Luciocephalidae

Since pikeheads (Luciocephalidae) have a labyrinth organ, they were often classified in older aquarium literature as being labyrinth fishes. The labyrinth organ of pikeheads, however, is structurally much simpler than that of the other labyrinth fishes. It is less convoluted, there are no lamellae inside the organ, and the capillary system is not as well developed. The most important role of the labyrinth organ in pikeheads may be as an adjunct to the inner ear, serving as a sound amplifier—pikeheads are known to have very sensitive hearing. They also use the organ, as labyrinth fishes do, for the absorption of atmospheric oxygen, but unlike most of the labyrinth fishes, the pikeheads use this organ for breathing only when under stress, when kept at high temperatures, or in oxygen-poor water. There are vast differences in their skeletal structure compared to that of labyrinth fishes. The first ray in the pelvic fin of luciocephalids is a hard spine, whereas in labyrinth fishes all pectoral rays are soft rays. In some aquarium literature it is reported that pikeheads do not have an air bladder. This is not correct. It is clearly evident from Liem's studies that not only do they have an air bladder but the air bladder is similar in shape to that of labyrinth fishes. *Luciocephalus* is a monotypic genus.

Luciocephalus pulcher (Gray, 1830/34)
SYNONYM: *Diploterus pulcher* Gray, 1830.
COMMON NAME: Pikehead.
LENGTH: Up to 7 in. (18 cm).
FIN RAY AND SCALE COUNT: D, 9–12; A, 18–19, P, 15–16; Pel, I/5; L.l. scales, 40–42.
DISTRIBUTION: Pikeheads occur on the Malay Peninsula, on Sumatra, Bangka, Belitung and Borneo. They are also reported from some islands of the Riouw Archipelago south of the Malay Peninsula. They were first imported in 1905.
DESCRIPTION: The body is elongate and oval in cross-section. The head is long and pointed. The fish has a very deep jaw gape and a highly protractile mouth. Several inter-connected jawbones can be protruded far forward to form a funnel-like extension of the mouth. Small fishes and other organisms are literally sucked in through this funnel. The lower jaw extends out beyond the upper jaw. The dorsal fin is small and is attached far to the rear of the back. The anal fin is deeply cleft in the center, so that on a superficial glance it appears that the fish has two anal fins. The caudal fin is small and rounded. The front hard ray of the pectoral fins is elongated and threadlike.

The background color is yellowish brown to brown. A stripe along the center of the body begins at the snout and runs through the eye to the caudal peduncle. Parallel to this stripe and beneath it is a second stripe consisting only of small spots somewhat separated from each other. The back is dark brown, and the belly a light yellow to cream color. There are considerable color differences between individual pikeheads, depending on their place of origin.

The pikehead has the typical shape of a highly predatory fish. It is capable of tremendous bursts of speed as it lunges from a concealed position in weed beds at any prey that swims by. However, typical of such predators, it does not have the endurance for prolonged swimming, which is usually not necessary anyhow, because of its relatively sedentary mode of life.
MAINTENANCE AND BREEDING: In older literature it is stated that this species gives birth to live young. This information is attributed to the first importer of pikeheads, who found small, young fish in the transport water. Certainly the divided anal fin which looks something like a copulatory organ could have given wings to this importer's imagination. Today it is known that the pikehead is a mouthbrooder.

In captivity pikeheads are very difficult to keep alive. They do best in a single species aquarium containing soft acidic water. This species requires strong filtration in the aquarium, such that there is a strong constant water current. This is best accomplished by using a motor-driven power filter, rather than an air-driven

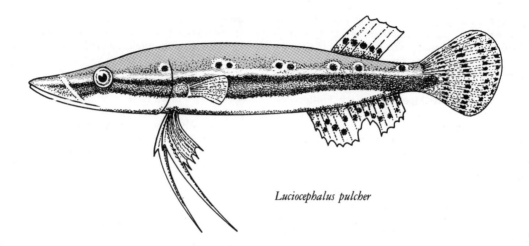

Luciocephalus pulcher

filter. Pikeheads should be kept in subdued lighting. Plenty of hiding places should be provided. Pikeheads only accept live food but are not particular about what kind; anything that moves will be eaten: large daphnia, mosquito larvae, dragon fly larvae, smaller food fishes, and flying insects knocked onto the surface. They can be trained to accept earthworms. Even with the best care, pikeheads rarely live long in an aquarium. If all of their environmental requirements are met, the cause of these usually premature deaths is probably to be found in their nutrition. In nature these fish probably live mainly on flying insects, which they catch at or directly above the water surface. The best success in keeping them has come from feeding them small crickets and live fish such as feeder guppies.

Snakeheads

Family Channidae

The fish discussed so far belong to the perchlike fishes, the order *Perciformes*. The snakeheads (family *Channidae*), which, like the labyrinth fishes, are also equipped with an additional breathing organ, represent a separate order, the *Channiformes*. Frequently they are also designated as the *Ophiocephaliformes*. Formerly, snakeheads were divided into two genera, *Channa* and *Ophiocephalus*. Today, all snakeheads are considered to be of the genus *Channa*. Fossil finds of these fishes are known from the upper Tertiary (Pliocene) period.

The snakehead body is elongated, and in cross-section the front and center parts of the body are almost round. Toward the caudal peduncle the body is some-

what compressed on the sides. The scales are of two types: a small, round, smooth type called cycloid scales, and a comb or toothed type, called ctenoid scales. The head is large and resembles that of a snake, especially when one looks at the very large scales on the top of the head. The mouth is deeply cleft, shows a complete set of teeth, and can be stretched very far open. The nostrils are located at the tips of tube-like projections. These fishes have no spinous fin rays. The dorsal and anal fins are very long-based, the caudal fin is rounded, the pectoral fins are attached far down on the side of the body, and the base of the pelvic fins clearly lies posterior to the attachment of the pectoral fins. One species, *Channa asiatica*, has no pectoral fins. In some species individual specimens as well as entire populations have no pelvic fins. The presence or absence of pelvic fins is thus not a generic characteristic. According to recent taxonomic works the genera *Channa* and *Ophiocephalus*, previously separated by this character, are now brought together in the older genus *Channa*, which was established in 1777 by Scopoli.

In the snakehead fishes the breathing organ, which corresponds to the labyrinth organ of the labyrinth fishes and pikeheads, is located in a hollow space above the gill cavity. This organ is relatively large, abundantly convoluted, and very effective in its function. It enables the fish to live in oxygen-poor water. Tropical species can withstand water temperatures up to 104°F (40°C) without harm. The Amur snakehead, coming from the temperate climate zone, has been released in places in southeast Europe, where it is able to live in waters in which even tenches and crucian carp, which are known for their ability to survive in polluted water, cannot survive. The snakeheads are

Descriptions of Species

very adept at moving overland, doing so by slithering like a snake and aiding their movements with their pectoral fins. Some snakeheads survive short drought periods by burrowing into moist bottom mud.

Their distribution extends in Asia from the Amur area to Korea and China and to Southeast Asia, including the Malay Archipelago and the Philippines. Other species occur in Africa. Because of their significance as food fishes, snakeheads were released a long time ago in waters outside their original range. For this reason, the original distribution can no longer be determined for most species.

Snakeheads are predatory fishes, devouring everything they can swallow. They eat fish up to ⅔ of their own length. Whereas they hunt smaller fishes rather aimlessly, they stalk larger fishes quite carefully. Before attacking, they bend their body in an S-shape curve, and then lunge at their prey. Larger fishes are turned around after having been grasped and are swallowed head first. They also eat other aquatic animals, including amphibians. Larger specimens can even catch small water birds on the surface.

During spawning, snakeheads embrace in a similar manner to that of labyrinth fishes. The male wraps himself around the female so that their genital openings lie directly next to each other. The floating eggs, each containing a large oil droplet, rise to the surface, and generally are guarded there by the male until they hatch. Nest-building so far has been observed only in the Amur Snakehead (*Channa argus warpachowskii*) which builds a nest from plant parts. Large species are reported to lay up to 50,000 eggs. The eggs hatch after 35 to 40 hours and the larvae first drift belly-up at the surface. They start swimming after another three to five days, often before their yolk sac has been completely absorbed. According to recent findings, there are mouthbrooders among the snakeheads. In repeated breedings of *Channa orientalis,* it was determined beyond question that this species is a mouthbrooder. The eggs are taken into the mouth by the male and remain there until their development is completed. The female participates actively in the brood care by defending the territory. She does not seem to take up the eggs, however. According to available reports, the young are not released through the mouth but through the gill openings.

If small snakehead species are kept alone or with larger robust fishes, they will thrive well and may even breed. Individuals of the same species and same size get along peacefully. Snakeheads are lively active fishes. After some time in an aquarium they will accept pieces of food directly from the hand. It is not necessary to feed them live fish all the time; imported snakeheads usually quickly learn to eat dead food such as fish flesh,

beef heart, and liver. Live foods, such as mosquito larvae, earthworms, tadpoles, mealworms, and crickets are readily accepted. Since they are gluttonous eaters, they produce a lot of waste, so frequent partial water changes and good filtering are called for. The aquarium should be tightly covered since all snakehead species are known to be good jumpers. If a half dried-out snakehead is found on the floor, put it into a very shallow platter of water; usually it will survive the incident. The larger species can be bred in large aquariums. The sexes can be determined during the spawning period by the more vivid colors of the male and the larger girth of the female. Aquariums of at least 25 gallons (200 liters) are advisable. The fish usually spawn after a large volume of the water is changed and at a temperature of 77 to 86°F (25 to 30°C). The young fish can be fed immediately with brine shrimp nauplii when they begin to swim on the sixth or seventh day after spawning. Snakeheads are extremely fecund, but since young fish, like their parents, have cannibalistic tendencies and immediately eat their smaller siblings, over-production of young rarely is a problem. The growing young fish often show schooling behavior. Of the over thirty species that have been described to date, only those species that have been frequently imported will be considered in the following species descriptions.

Asian Species

Channa lucia Cuvier and Valenciennes, 1831
LENGTH: Up to 16 in. (40 cm), usually smaller.
FIN RAY AND SCALE COUNT: D, 39–43; A, 27–30; L.l scales, 58–65.
DISTRIBUTION: Thailand, Cambodia, Vietnam, Sumatra, Borneo, and Java, in all types of water, abundantly in some places.
DESCRIPTION: Primary coloration is brownish, darker on the back, changing to yellow toward the belly. Two small horizontal stripes run along the center of the body and are connected by an irregular spotted pattern.

Channa marulia (Hamilton and Buchanan, 1822)
COMMON NAME: Indian Snakehead
SYNONYMS: *Ophiocephalus marulius* Hamilton and Buchanan, 1822; *Ophiocephalus leucopunctatus* Sykes, 1841.
LENGTH: Up to 4 ft (120 cm), the largest of all the snakeheads.

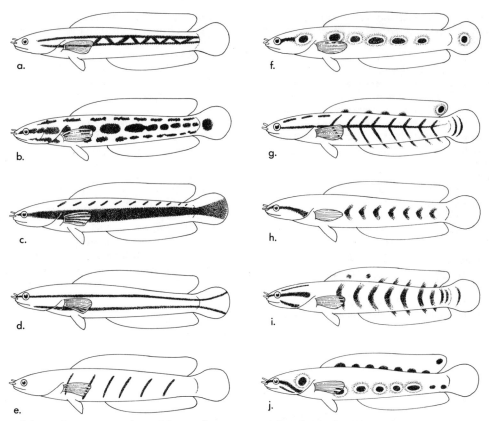

Schematic representation of the marking pattern of snakeheads. The marking patterns which are usually very pronounced in young fish fade in older specimens. a = *Channa lucia*; b = *Channa marulia*; c = *Channa melasoma*; d = *Channa micropeltes* e = *Channa orientalis*; f = *Channa pleuropthalma*; g = *Channa punctata*; h = *Channa striata*; i = *Channa africana*; j = *Channa obscura*.

FIN RAY AND SCALE COUNT: D, 49–55; A, 28–36, L.l. scales, 60–70.

DISTRIBUTION: From southern China, over the large part of peninsular and continental India, to the islands of Sumatra, Borneo, and Java, and in the Sunda Archipelago.

DESCRIPTION: Young fish have two dark horizontal stripes along the center of the body and an orange band between these stripes. This pattern later turns pale and is replaced by a dark brown background color against which three rows of dark spots are clearly contrasted. The dorsal and anal fins also have a pattern of smaller brown spots. This species can be kept only temporarily as a young fish in a home aquarium. Adult specimens are excellent show fish for public aquariums.

Channa melasoma (Bleeker, 1851)
COMMON NAME: Black Snakehead
SYNONYM: *Ophicephalus melasoma* Bleeker, 1851.
LENGTH: Up to 14 in. (35 cm).
FIN RAY AND SCALE COUNT: D, 37–40; A, 22–25; L.l. scales, 40–55.
DISTRIBUTION: Vietnam, Thailand, Malay Peninsula, Sumatra and Borneo.
DESCRIPTION: The background color is brownish green to greenish blue, with a red horizontal stripe along the center of the body. The misleading name "black snakehead" can be explained by the fact that specimens preserved in alcohol take on a very dark coloration. Unfortunately, young fish of this species are rarely imported, which is too bad since they mature when half grown and would be suitable for breeding in a home aquarium.

Channa micropeltes (Cuvier and Valenciennes, 1831)
SYNONYM: *Ophiocephalus micropeltes* Cuvier and Valenciennes, 1831.
LENGTH: Up to 1 yd. (95 cm).
DISTRIBUTION: India, Burma, Thailand, Cambodia, Vietnam, Malay Peninsula, Borneo, Sumatra, and Java.
DESCRIPTION: In nature this species is a fish-eater. The frequently imported young fish have bright markings. Two dark horizontal stripes run alongside the center of the body on an orange background. The upper stripe extends from the tip of the snout through the eye into the upper part of the caudal fin; the stripe below it extends from about the throat to the lower part of the caudal fin. Adult fish have a greenish gray background color with light gray spotted markings and a light gray belly. Unfortunately the prettily colored young fish grow quickly, so that they can be kept in a home aquarium for only a short time.

Channa orientalis Bloch and Schneider, 1801
SYNONYM: *Ophicephalus gachua* Hamilton and Buchanan, 1822.
LENGTH: Up to 12 in. (30 cm), but usually smaller. Regular dwarf populations from mountain brooks are known.
FIN RAY AND SCALE COUNT: D, 34–37; A, 21–23; L.l. scales, 41–45.
DISTRIBUTION: From Afghanistan, across the entire Indian subcontinent, including the Sunda Islands. It also occurs on Sri Lanka.
DESCRIPTION: The background color is yellowish brown to dark brown. Young fish are often greenish brown. There are considerable variations in color and markings within its huge geographical range and the very different biotopes in which this species occurs. As far as is known, most populations seem to have a red edging on the unpaired fins. On Sri Lanka there is a population that completely lacks pelvic fins. This species has been imported frequently and successfully bred.
MAINTENANCE AND BREEDING: This species is a mouth breeder, the male being the brooding parent. Whether this species is generally mouthbrooding or whether this habit is valid only for populations from certain extreme biotopes has not yet been investigated.

Channa pleurophthalma (Bleeker, 1850)
COMMON NAME: Spotted Snakehead
SYNONYM: *Ophiocephalus pleurophthalmus* Bleeker, 1850
LENGTH: Up to 14 in. 35 cm).
FIN RAY AND SCALE COUNT: D, 40–43; A, 28–31; L.l. scales, 57–58.
DISTRIBUTION: Sumatra and Borneo.
DESCRIPTION: The background color in the upper half of the body is brown, changing to yellow toward the belly. On the sides occur six light-edged dark spots.

The most prominent of these spots is on the gill cover, and the rearmost is on the caudal peduncle. The distance between the first five of these spots is equal, and that between the fifth and sixth spots is larger. The dorsal and ventral fins posteriorly have a pattern of dark brown, halfmoon-shaped stripes. The species is frequently imported and has been bred.

Channa punctata (Bloch 1793)
COMMON NAME: Dotted Snakehead
LENGTH: Up to 14 in. (35 cm).
FIN RAY AND SCALE COUNT: D, 29–32; A, 20–23; L.l. scales, 37–40.
DISTRIBUTION: India and Sri Lanka. This species was first imported in 1893.
DESCRIPTION: The background color on the upper side is light brown. On the back is an iridescent color. The sides of the body are more yellow brown to bluish yellow and the belly light gray to almost white. Younger fish have a black horizontal stripe along the center of the body which can barely be seen in adult fish. Adult fish show a pattern of spots and slanted stripes. On the rear end of the dorsal fin there frequently is a clearly defined dark spot. The dotted snakehead is a significant food fish in its homeland. This species has been bred in captivity.

Channa striata (Bloch 1793)
COMMON NAME: Horizontally striped snakehead
LENGTH: Up to 35 in. (90 cm), but fish of this size are rare in open waters.
FIN RAY AND SCALE COUNT: D, 38–43; A, 23–27; L.l. scales, 52–57.
DSISTRIBUTION: From southern China across the entire Indian subcontinent to the Sunda Islands. It also occurs on the Philippines, but it was probably introduced there by man.
DESCRIPTION: The background color of the upper half of the body is brownish gray to greenish gray; the lower half of the body is yellowish gray. Young fish show a dark horizontal stripe along the center of the body. Oblique vertical stripes form a herringbone pattern between the stripes. This marking fades in older fish to only a few dark spots along the base of the dorsal fin.

African Species
Channa africana (Steindachner, 1879)
COMMON NAME: African Snakehead
SYNONYM: *Ophicephalus africanus* Steindachner, 1879.
LENGTH: Up to 13 in. (32 cm).
FIN RAY AND SCALE COUNT: D, 42–49; A, 30–34; L.l. scales, 74–82.

DISTRIBUTION: West Africa, from Nigeria to Cameroon.

DESCRIPTION: The background color is loam brown to
light gray. Adult fish occasionally show a pattern of
chevronlike vertical stripes from the center toward the
back. There is a dark spot on the gill cover from which
a short stripe runs to the eye. The fins are light grayish
brown, and the dorsal and anal fins are adorned with
small spots and lines. This species is rarely imported
and nothing is known about its breeding in captivity.

Channa obscura (Günther, 1861)
SYNONYM: *Ophicephalus obscurus* Günther 1861.
LENGTH: Up to 14 in. (35 cm).
FIN RAY AND SCALE COUNT: D, 40–45; A, 26–31;
L.l scales, 62–76.
DISTRIBUTION: Across wide areas of Africa, from the
river basin of the White Nile to the Chad Basin to
the Senegal River. It occurs southward all the way to
the Congo River basin.

DESCRIPTION: With its pretty markings, it is probably
the most frequently kept snakehead. The back is olive-
brown, which becomes lighter brown farther down the
sides. Along the center of the body is a series of elongated spots forming a horizontal stripe that runs from
the tip of the snout to the caudal peduncle. Above
and below the horizontal stripe numerous dark brown
to black lines and spots form a pattern that reaches to
the top and bottom of the dorsal and anal fins. During
the spawning period all brown spots darken and contrast strongly against the lighter background. This
species is easily kept in an aquarium. Since it becomes
sexually mature at a body length of 6 in. (15 cm), it
is well suited for aquarium breeding. It has been bred
frequently.

INDEX

Index

Index